Every Fish Tells a Story

Reef Society in Comedy and Tragedy With Hope for the Future, Maybe

Snorkel Bob

Skyhorse Publishing

Skyhorse Publishing books may be purchased in bulk at special discounts for sales promotion, corporate gifts, fund-raising, or educational purposes. Special editions can also be created to specifications. For details, contact the Special Sales Department, Skyhorse Publishing, 307 West 36th Street, 11th Floor, New York, NY 10018 or info@skyhorsepublishing.com.

www.skyhorsepublishing.com

10 9 8 7 6 5 4 3 2 1

Library of Congress Cataloging-in-Publication Data is available on file.
ISBN: 978-1-61608-396-0

Printed in China

Special thanks to United States of America Secretary of State Hillary Rodham Clinton for reminding us:

It takes a reef to raise a fish.

To our 41st President of the United States of America Ronald W. Reagan, who let freedom ring with:

Mr. Gorbachev, tear down *this aquarium!*

And to Rod Stewart for his original classic that still plays in aging boomer heads across the Cosmos:

Every fish here tells a story story.
Every fish here tells a story.

Sleek unicornfish, left, and saddle wrasse review today's potential: *I don't know, Marty. Whadda you wanna do?*

Ed Lindsey

Many dedicated reef huggers spoke up during Maui's campaign against gill nets, walls of death that kill every ocean creature they touch—fish, monk seals, turtles. Chief among the front line troops was my friend Ed Lindsey, a Hawaiian and charismatic leader who died not too long ago but whose voice still echoes. During the contentious gill net struggle, with claims of cultural heritage abounding, Ed bellowed over a rowdy crowd at a State-run forum:

"You should be *ashamed* of yourselves!"

A pin dropped.

In the silence, Ed recalled the *Kumulipo*, the Hawaiian chant tracing life on earth to its reef origin, the coral polyp.

Maui alone in the Hawaiian archipelago has a 100% ban on gill nets.

Following that historic hearing I asked Ed how he could say such things and have everybody think he's a great guy, whereas they'd want to strangle me, Snorkel Bob, for yelling at them. Ed reflected and agreed, "Yeah, you better not try that yet." Ed Lindsey had cancer for five years, campaigning full time to save what he could. He weakened and passed in short order, though his light still shines. I think my favorite image of Ed came a few days before he died, as he rested at home on a breezy afternoon, watching the gentle break over the reef behind his house in Lahaina.

Up the road at Kamehameha School where Ed and Puanani taught, the kids asked why Ed wasn't coming around. Puanani explained as delicately as she could, and that was that, till the kids asked the school bus driver to stop at Uncle Ed's house that afternoon, where they filed out and around to the ocean side where Ed could watch them line up and dance a hula and chant.

This is for Ed Lindsey, *Kanaka o'Kekai*.

The night gave birth

Born was Kumulipo in the night, a male

Born was Po'ele in the night, a female

Born was the coral polyp, born was the coral, came forth

Every Fish Tells a Story

—not just a personal story, but a bigger story too.

I laughed.

I cried.

It changed my life.

Aloha kakou,

If heaven exists on earth—that is, a place available in this life where behavior is purely motivated, where color and pattern are garish and balanced—then a reef is that place. For all we know, the fishes are us; we are them, rounding the Wheel of Life on different spokes. Most humans feel naturally superior to other species, even as the others demonstrate greater native intelligence, skill sets, common sense and compassion.

Every fish tells a story, some epic, some anecdotal. A few of those stories are recorded here, because the world's reefs are in peril, and if we listen to a species other than our own, we may learn. Many reefs vanished in the last few years and will not be soon or easily replaced. Reefs are deeply loved in many places, though squandered in other places. The companion edition to this one, *Some Fishes I Have Known,* is based on the premise that fish are social, sentient beings who see, feel, and know, with photos to prove the concept. "Proof" may be a stickler to the more data retentive among you, and for that reason *Some Fishes* stipulates that reef assessment can occur anytime anyone stares at fish while sucking air through a plastic tube. Advanced degrees are not required. The eyeball/brainpan synapse can be trusted. I snorkel, therefore I see.

The photos in this 2nd erudition on fish society also illustrate what any observer might *see* on opening her eyes, heart, and mind. That is, the cold and scaly gill-breathers of outrageous coloration and unbridled innocence are often willing to engage us. To take things a step further, many reef critters are as individual in character as some humans. Reef life occurs in a reef community, in a social order that differs from the open ocean world of game fish. Different cultures do not preclude individual character in either arena. Open ocean game fish take color and beauty to unimaginable heights, if you see them in person, in the water. If you're lucky, they see you back as few others will, for who you might be. Like during a swim-stop on board Sea Shepherd's Ocean Warrior* during the Eastern Caribbean campaign to stop Japanese whaling. We were 20 miles off the island of Hispaniola in the Caribbean Sea, where visibility is hard to judge as you scan the depths and wonder if that's 100 feet or 300. I followed a sunbeam as deep as I could till looking up, amazed and proud of the 40-foot ascent above. Soon a gang of rowdy mahi mahi swam over to our side of the vessel, where a 5-foot bull eased alongside yours, Snorkel Bob's, truly, shoulder to shoulder and eye to eye to ask in so many non-words: *You want to school up and run with us?*

Oh, Neptune, did I ever! Which is not to say that every mahi wants to be my pal, but some do. That's the point here: science-minded journals based on "data" might profile a species in terms of physical detail but won't tell you what a fish is about. Are humans finicky? Fearful? Shy? Bold? Cruel? Compassionate? Able to love nature? The most deadly of species? * See my Snorkel Bob's log, *On Board the Ocean Warrior* at www.snorkelbob.com.

Well, yes, but assigning traits to an entire species is crass. Look at these 2 species, one mammalian, the other eelian. It's plain to see 2 individuals in "attack" mode here, jaws agape, fangs bared, bloodlust apparent, and so we judge them:

Kukla the eel, blood-lusty killer of the deep.

Rocky: Where's the beef?

Yet given a moment or 2, the jugular ferocity we plainly saw is not what either individual intended. These guys are waking up from nappy poo!

Uh, do you have any squibs? I mean squids?

Can you get the lid off this thing?

Which is not to say that all sea creatures are as cute and cuddly as Kukla and Rocky. They're not. Some are aggressive, like Gargantua (right), who would like to eat you but can't, because he's hardly bigger than Rocky's dingdong.

Or Lizardfish, who knows his limitations but presses slyly on:

Well hello, little fish. Would you like to stay for lunch?

These pages are meant to entertain, to tell a few stories from the fishy P.O.V. in a context of neighborhoods, of individual homes to specific personalities who might live there for years with continuing relationships (friendships?) among neighbors and visitors, with you and me on the potential guest list. Sadly, the overwhelming story is one of pressure from human population growth. Its toxic effluent and need are killing nature faster than nature's ability to recover.

I'll try to keep the grim stuff brief—but if you turn these pages, you likely care what happens, and you should know these things. These stories may help you see what's going on in the seas around us. Science often resides in the eye of the beholder, with data spun as wickedly as political nuance, to favor a point of view. We must assess data in the context of who gathered it and the objective of the sponsor.

Name calling isn't productive, but crime against nature is worse. Reef Check International and its many branches ask for money so you "can make a lifetime of difference for our coral reefs!" Reef Check promotes the idea that monitoring reef health will restore reefs. But among Reef Check's directors is the biggest reseller of Hawaii reef fish in the aquarium industry, a fellow calling himself a stakeholder in Hawaii reefs as he profiles me, Snorkel Bob, as having "no credibility in the industry."

That's like saying Simon Wiesenthal has no credibility with the skinheads.

How many millions of fish leave Hawaii each year for the aquarium trade? Nobody knows, though the biggest exporter estimates $20 million gross accruing to Hawaii's economy with an average of $4-5 per fish. Can you divide 20 million by 4 or 5?

Yellow tangs are 60% of the take—tangs are herbivores who live to 40 years on a reef, eating algae dawn to dusk, yet they die in year or 2 in a tank. Many species critical to reef maintenance now ship to the US, Europe and Asia as aquarium "livestock" from Hawaii, where reefs are collapsing from algae overgrowth.

Aquarium collecting is emptying Hawaii reefs, and everyone sees it. Reef Check pitched Snorkel Bob's a few years ago on coloring books for kids; we could get a great order for only $50,000. The product looked to me like a load for the landfill with no relevance to reef recovery and a wrong message to kids. We get pitched often on bad ideas, yet this one lingered; who are those guys? How can Reef Check come on so strong with so little substance? Why does a money-ferret travel with an entourage and know the particulars on maximum salary the IRS allows a non-profit director? The details filled in soon enough.

A big wildlife fund has a policy to protect wilderness so that wilderness can better serve humanity—but that's not wilderness; that's a wilderness theme park. That fund works with Reef Check. I could go on to The Nature Conservancy, a giant land-trust heavily directed by developers and bankers. Do we have values issues? The list is long, and you should know what they're up to—especially if they want to "protect" 3rd world areas with "sustainable" agendas just as missionaries came with the word of God. Will they export resources? Will they condone and then promote aquarium extraction? Did somebody say Conservation International?

I, Snorkel Bob, got a questionnaire from a coalition of HUGE non-profit agencies only yesterday, such as it is. Question 4 asked:

How many people or companies now working in the aquarium industry do you think would go along with a program to safeguard coral reef wildlife?

Answered I, Snorkel Bob:

All of them, as long as they can make it "sustainable" and continue taking everything in reach.

Hey, when it's down to the nitty gritty I, Snorkel Bob'm not known for coddling culprits. DAMN, these guys fry my oysters! But I'm calmer now. Just do me a favor: use common sense on the reef and off. When it comes to data, trust your own lying eyes. For fox* sake, do we actually need people with advanced degrees and vested agendas to validate the reality before us?

No, we don't. And on that note we delve again into an extraordinary social order peopled with individuals who happen to be fish.

Yours in the bond,

Snorkel Bob, Himself

Snorkel Bob, Himself

* small, red, intelligent animals deeply in need of our love, compassion, and care. Chasing them to death on horses with dogs out front is not nice. It's vulgar.

Oh, God. What time is it?

This flounder tells a story. Maybe you recognize the face or the feeling. Where do flounders go at night to carouse? How deep are their vices? Are they taciturn and morose most mornings?

That's a spin too, and a better question might be: who doesn't cu… ccc… cccu... ccccut her loose (!) every now and again?

Here's our friend a short while later looking better already, only a latte away from optimal.

Did you say something?

Okay, let us begin on a chipper note, as prelude to the heavy stuff, with a heart-warmer surely to engage, at our starting point below the line a long way from home in thc land of…

Oz & The Great Reef

Where life comes at you in waves, and the ending has yet to be written...

I call him Baoota (bah•OOta)—so big, hunky, fearless and friendly. He's a humphead wrasse, also known as Napoleon wrasse or Maori wrasse, and like most wrasses, his curiosity makes him a character to reckon on the reef. I, Snorkel Bob, read a fish-guide profile calling the Maori wrasse wary & reclusive, but individual fish often vary from their species or group in personality traits. Are all New Yorkers pushy? All Texans speech impaired? No, they're not, and nobody likes a crass generalization—oh, I'll hear about these 2.

Let's get past the bad news: a recent report on collapsing fisheries showed big bins of Maori wrasse rolling into an Asian fish market. These biggest of all wrasses live about 20 years and can grow to 7 feet, but the average size is down to 20 pounds, so more fish are needed to feed the growing hunger. The report further noted that the decline in fisheries is so great that many Asian restaurants now feature smaller reef fish to compensate for size with color and pretty shapes. They're often in aquariums, so you can pick the one you want to eat, like an ogre under a bridge with a cooking staff.

A boat captain on the Great Barrier Reef confided that he works for the Australian government agency similar to the CIA, keeping an eye out for poachers. They come south from Asian ports and cannot be stopped, especially at night. Most Australians I met in reef tourism view the Great Barrier Reef as a done deal—gone in 50 years, no doubt about it, from poaching and warming—and oil tankers. While you may feel helpless you must rage into this dark night, declining purchases from guilty countries and speaking out too. Tough, yes, but without a line in the sand, we go out with a whimper. Or is that a yawn?

Another captain told of carnage at Flynn's Reef when it got

opened to the aquarium trade for 3 days because of the "true" Nemo population there—that's the species of anemone clownfish actually shown in the Hollywood movie. That section of reef got closed in a day when a monitoring team found no "Nemo" remaining. The aquarium trade Hoovers every reef it can access, for the money, to feed its families, or some such sustainable euphemism for sheer, raw greed.

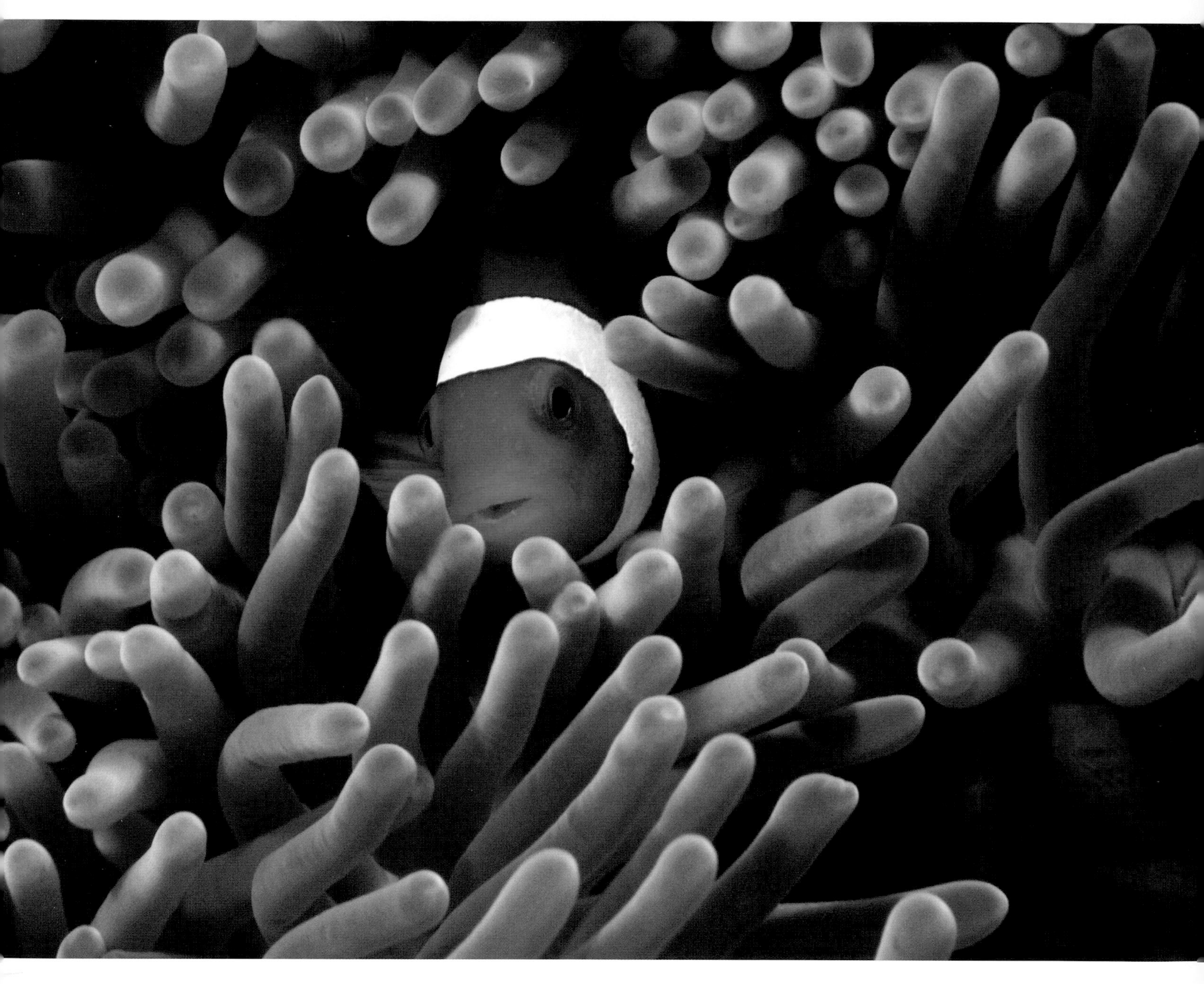

Another fellow in Cairns rents out point 'n shoot cameras in cheap underwater housings. He said 2 of his best mates are aquarium collectors and must follow more rules than a dog has fleas. He said most people don't like the aquarium trade, but that's only mass hysteria. It's like the resentment of his annual fishing trip to lush reefs, where he and the wife catch a thousand pounds of fish. Everyone gives him grief, but he cleans every fish himself before flash freezing—not a cheap unit, the flash freezer, and neither is the babysitter for that matter; ha!—and he and the wife and kids eat every morsel of that fish.

This failure of logic dooms the earth's oceans. Must I, Snorkel Bob, multiply this self-righteous fisherman's habit by billions of hungry humans?

Meanwhile, Baoota is neither wary nor reclusive but favors the direct approach, guaranteed to get your attention. Ordinarily, this obtrusive though gentle behavior would indicate tourists with fish food—but this is 40 miles offshore, 60 feet deep. The Great Reef is thoroughly managed, with moorings owned by specific companies and valued in the millions; so great is the demand for fish communion. Maybe Baoota became socialized. He sniffed my nose before turning sideways for a better view. Then he schooled up, enjoying the company and pointing out some key features of the neighborhood. I have rarely met a better reef host.

This cheeklined Maori wrasse below, on the other hand, is hardly 7 inches and weighed in about equal to one of Baoota's pectoral fins. He couldn't be bothered, other than to pose, bare a little tooth and cast a wary eye on interlopers, just as his fish-guide profile predicted.

Note the soft coral movement, indicating current and/or surge. Water movement seems prevalent here and can factor in navigation or planning. Since the fishiest reefs in the world are noted for strong currents, most fish huggers view moving water as a delivery system, bringing food to the fish minions. On the navigational aspect, I, Snorkel Bob, do not excel. In rough vernacular, I can't find my ass with both hands. This can hinder a shutterbug who wants to hang out, to work with a fish or 2, to make like flotsam and get familiar, to capture the sweetness, curiosity and concern.

Ah, well. On the Great Reef every diver must carry a hotdog—a bright orange tube that inflates to about 3' x 5" for visibility in 5' pitching seas. The general rule is: if you lose your buddies, search for a minute or 2 and then surface. But getting lost is a drag and can shorten a dive, so I, Snorkel Bob, kept up with those who knew the way. The Great Reef can be murky.

In fact the 5-hour trip from Cairns begged the question: WTF? Imagine a 15-knot breeze under gunmetal skies with breaking waves frothy as dog jaw and equally hostile. But the crew remained cheerful (Cha ching, cha ching). I, Snorkel Bob, know tourism. Tourism is a friend of mine. This is tourism. I.e., get the bodies in the water, and it's on the books; make the deposit.

Most amusing were the 20-something boys who naturally love chaos. They clicked their heels and devoured another egg sal san as the more seasoned folk accessorized in green and grasped their little paper bags—bags that reflect Australian consciousness by dissolving in 3 seconds flat when wet. No, really. 3 seconds is plenty for a lunge and toss over if you hurry. Well, the last laugh is always best, unless it's between hurls along with the rowdy boys. I, Snorkel Bob, did not ho heave but thoroughly comprehended prospects.

Not to worry, reef moorings are sorted for lee water in all conditions, and in no time it was down to 4' and only 12 knots. What a cakewalk! Especially after gulping seawater just off the stern, stuffing my gob with regulator and going DOWN, under, as it were, where you find yourself in moderate murk after rising at 5 AM for a 4-hour boat ride the day after a 13-hour plane ride and not exactly wondering why but surely asking where is the stuff I saw on the website?

We didn't linger in the murk but finned to the reef for shallower water and somewhat improved visibility, where a directionally challenged photog could keep up while focusing fast on the passing parade, which sounds like a push and felt like it too, till realization struck in mere moments that here is life, color, mystique, society and balance in grand effusion, teeming as any town in Asia, Africa or Illinois but *without* the pollution and overpopulation.

Longnose filefish

Longnose filefish are considered rare, though I, Snorkel Bob, think their encounter may be the luck o' the draw on season, current, time, tide and Neptune's fancy. Like filefish everywhere, longnose are tentative in their approach and dubious on contact. But a bloated log with a camera can sometimes ease on in for the family portrait.

Far more numerous and no less demure, the reticulated dascyllus:

Reticulated dascyllus pups

Whether a fish snuggles in soft coral or cruises over a coral thicket may depend on mood or needs. The soft corals appear pliant and receptive as quick cover for the reticulated dascyllus on page 17 or these charcoal damsels.

Charcoal damsels

But hard coral can confound predators—and provide shelter to these turquoise chromis.

Chromis in a coral cluster

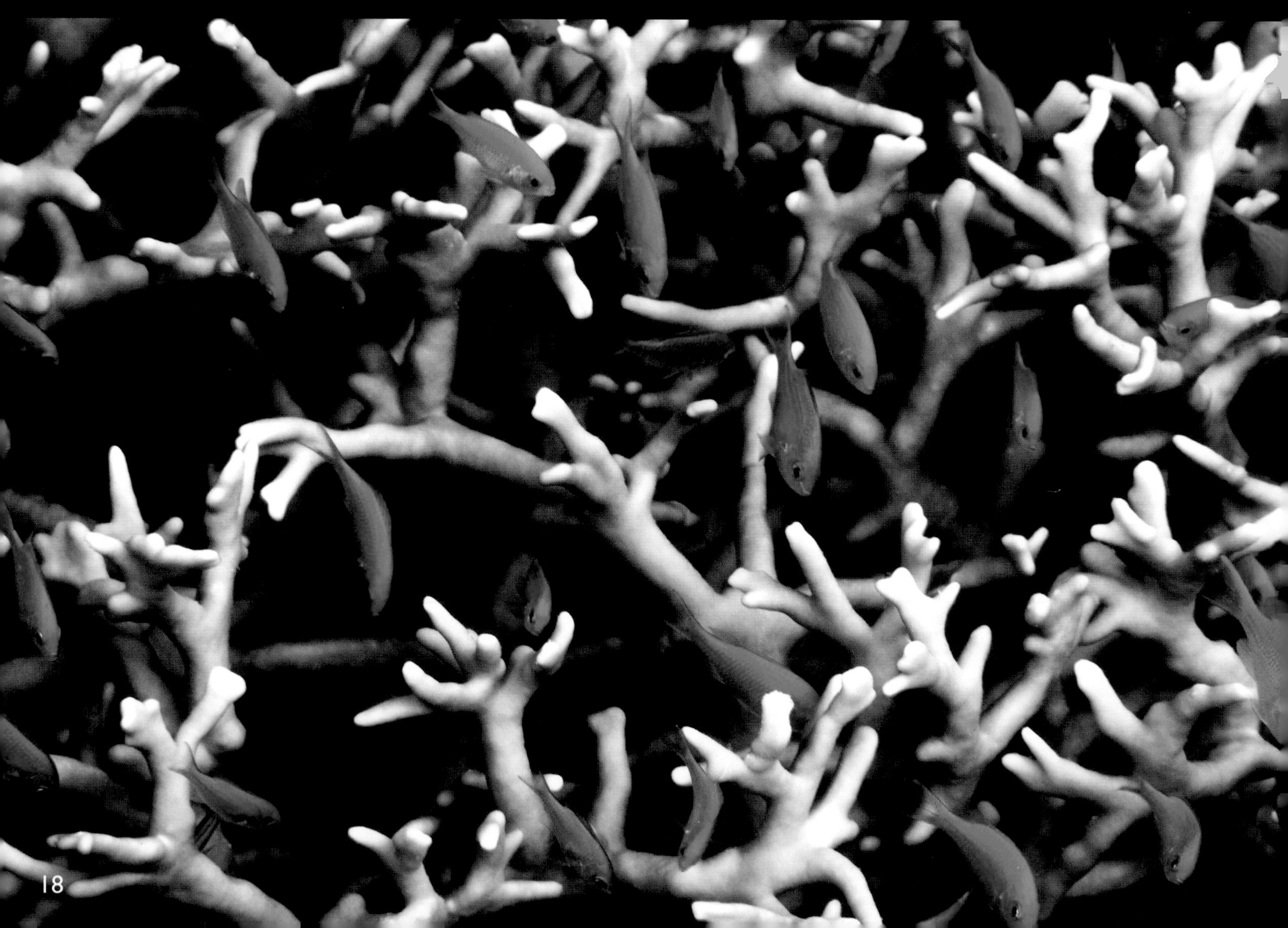

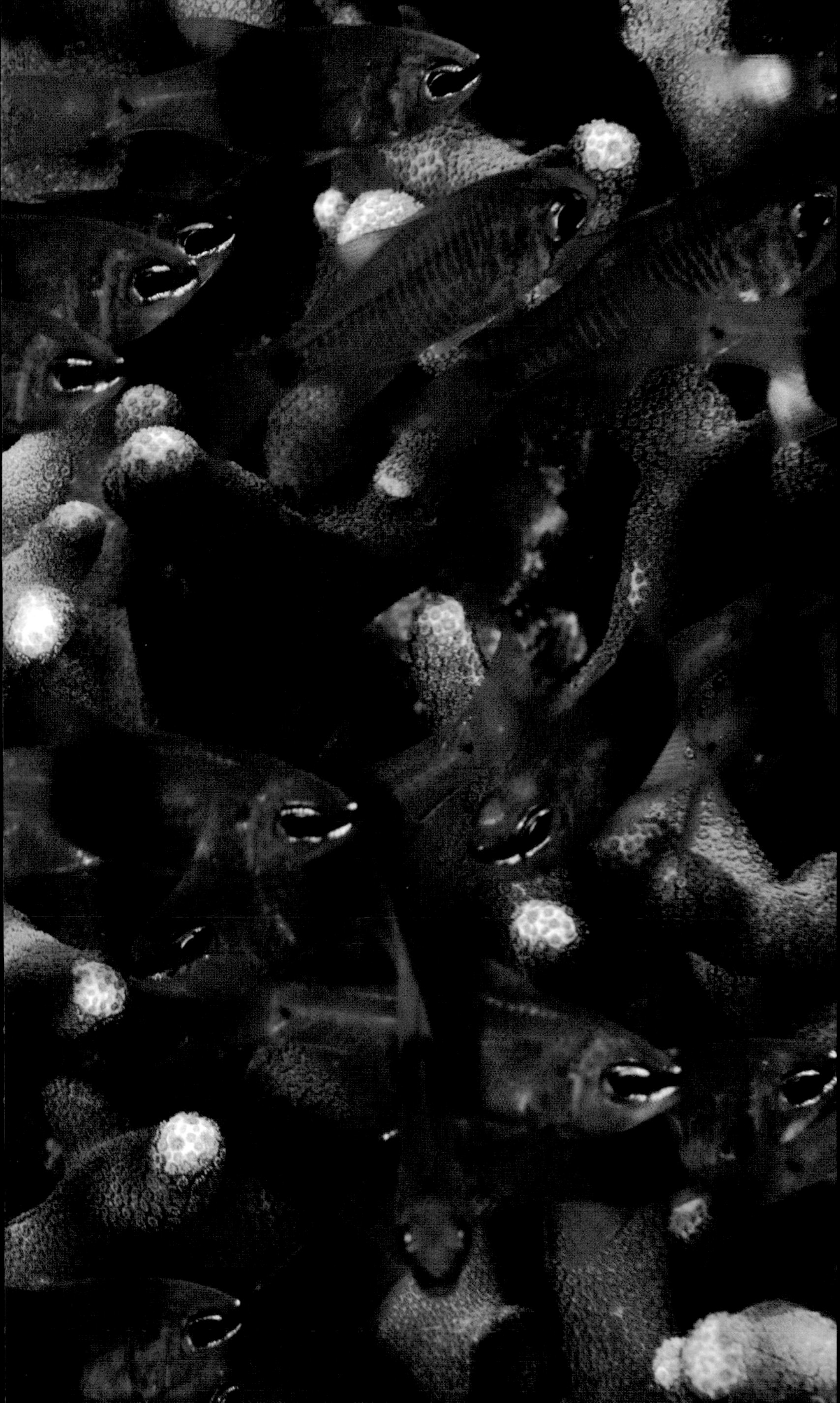

Coral cover is in the eye of the beholder. These two chromis species, narrow-lined and girdled, blend with the hard coral cover to give a predator vertigo.

Many fish maneuver among the corals for simple artistry, like these…

Tiny blue guy over soft coral

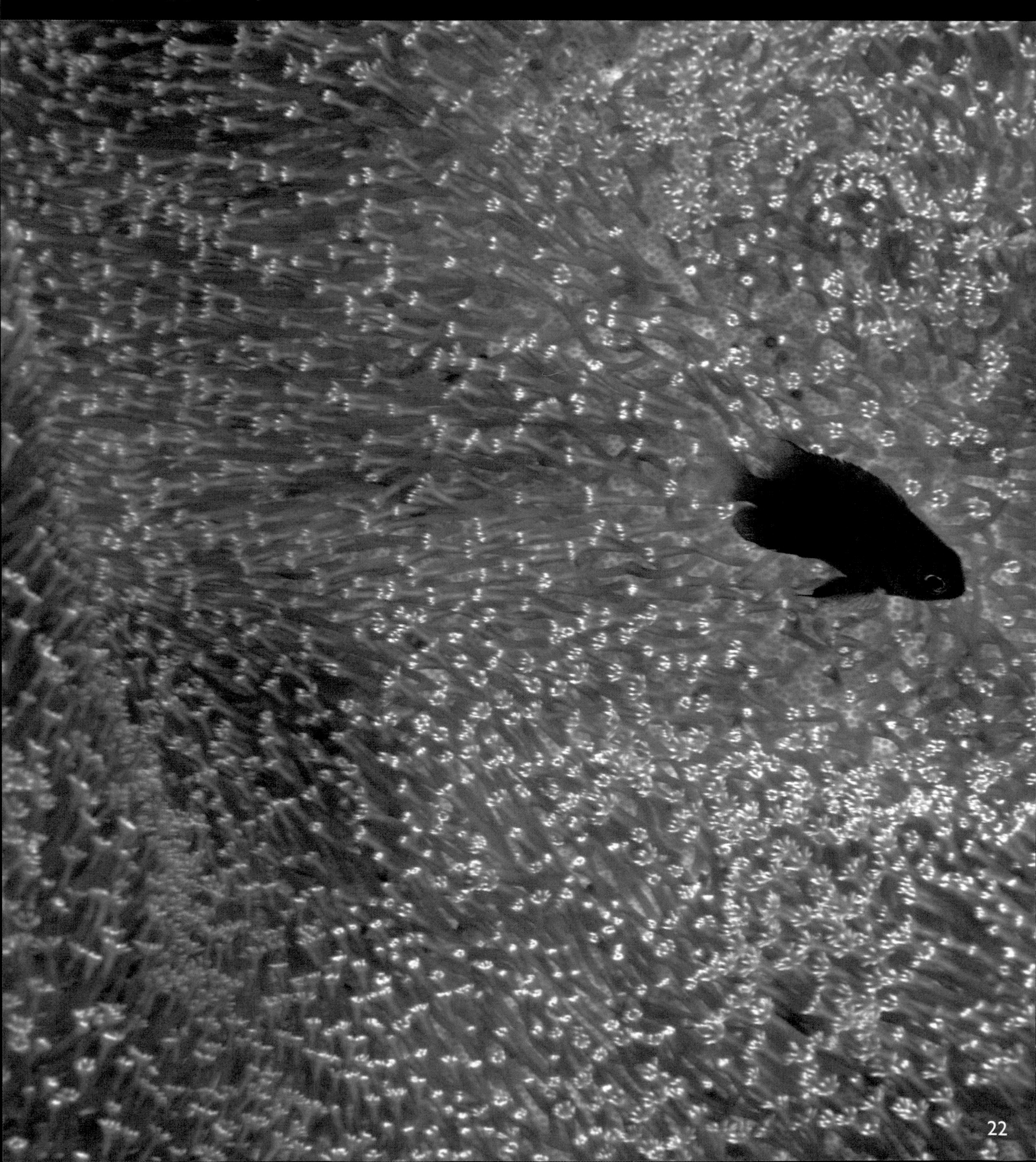

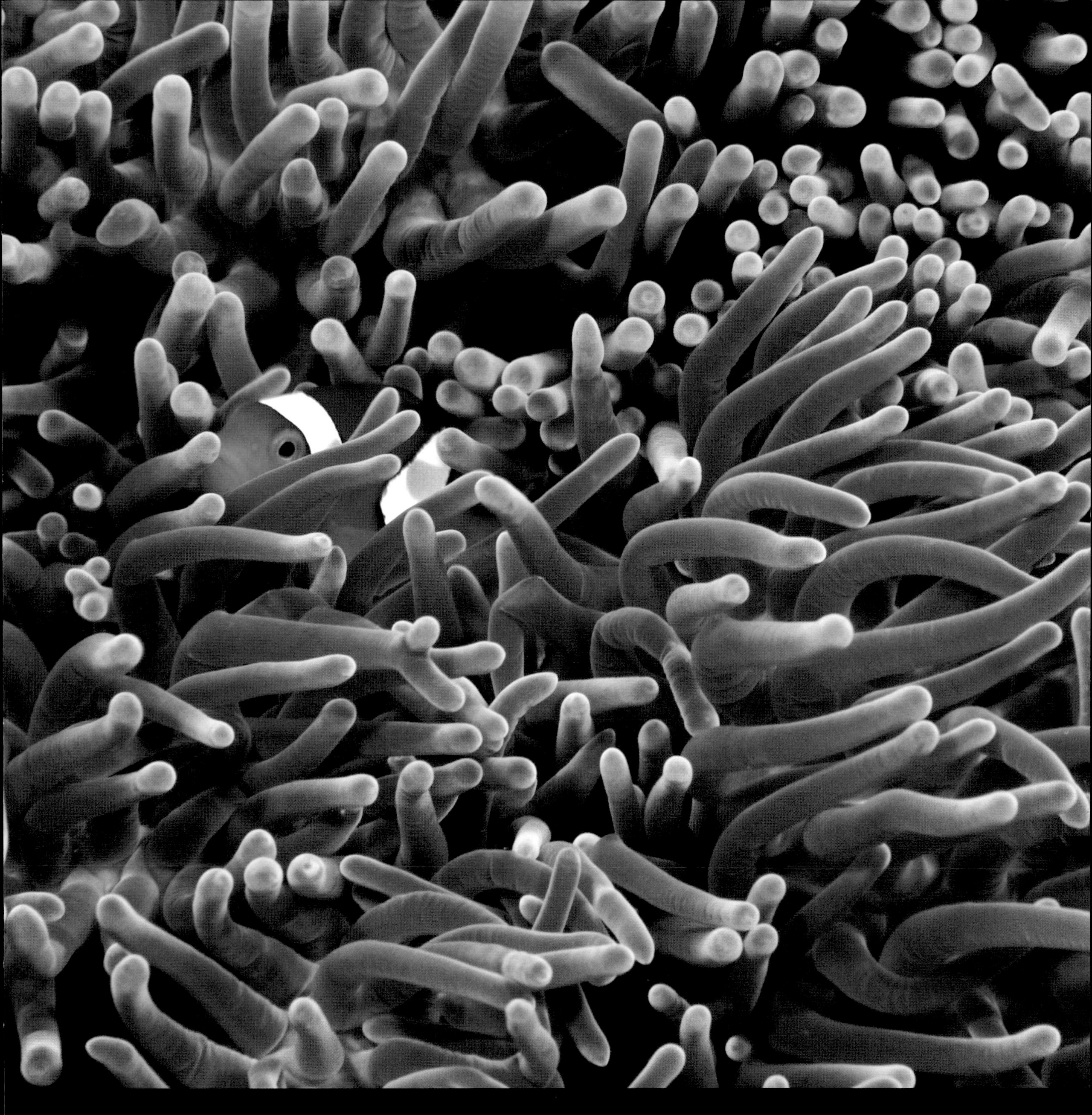

This is different. While the anemone might help hide the clownfish, the real protection is in stinging toxins in each swaying tentacle—the clownfish is immune. Like soft coral, the anemone is an animal, but it's one animal, not a colony of thousands, and it generally partners with clownfish or other damsels who live in its flowing folds. How does the symbiosis help the anemone? Clownfish makes the scene look comfy to other fish, who soon learn that dinner is on them.

Damsel flourish

Lemon damsel

Note: Damselfishes are a prolific species and include chromis and dascyllus, though the latter two are generally different in shape. Species identification skills come with encounters, and you'll soon develop an "educated"* eye for a fish's family or species. If you get it wrong, as I often do, you lose only a minute or two finding it in the fish ID books.

**University of the Reef, where data collection is unsponsored and weighted by common sense.*

Turquoise chromis in hard coral.

Turquoise chromis in soft coral.

A staghorn damsel on orange coral.

Dot and dash butterfly

Tiny torpedoes hanging out.

Little red at home.

A dartfish poses over soft coral.

Longnose filefish cruise for snax and cover.

The action comes at you on the Great Reef, with more fish than you'll see on most reefs anywhere in the world. Australian reef defense and management safeguard these fish, for now. Another safeguard is size—a reef this big is harder to manage but more likely to survive, for now. I, Snorkel Bob'm too often jaded by the Hawaii situation, where reef resources are claimed as a right by many user groups except for one, the fish—and the only state management in place is an agency charged with optimal revenue on fisheries.

WHICH MAY WELL BE GROUNDS TO INTERRUPT THIS PROGRAM for a brief story of my own. In recent years, meeting individually with ¾ of the Hawaii State Legislature in both House and Senate, the anti-aquarium league developed 3 cases to present, as applicable to any given legislator, group or pedestrian.

1) The economic case, comparing annual revenue of aquarium collecting in Hawaii, $2 million, which threatens reef-based tourism at $800 million. That's a 400-to-1 shot.

2) The cultural case, presenting *Kupuna* (elders) and *Kumu* (teachers) saying that aquarium collecting is an affront to Hawaiian culture.

3) The conservation case, that the aquarium trade is leaving Hawaii reefs empty with no limit on its catch, no limit on the number of catchers, and no constraint on endemic or rare species.

Blueline butterfly. "Experiencing a 100% decline (in Honaunau on the Kona Coast)..." That's Hawaii Department of Land & Natural Resources speak for: They're gone.

Or, in the vernacular: Shucks, we lost another one.

Or most concisely of all, the pidgin: Da kine, whodaguy, no mo stay. NOTE: The blueline butterfly is a relic species—not only endemic to Hawaii but with no other species similar anywhere in the world. Blueline butterflies are wholly unprotected from the aquarium trade.

THE CONSERVATION CASE HIGHLIGHTS charismatic endemics (found nowhere else in the world) leaving Hawaii daily, sure to perish in a month of captivity.

Hey, Cleaner Boy, I think you got a dingleberry.

The Hawaiian cleaner wrasse eats parasites from the skin and scales of other fish. Removing cleaner wrasses exposes a reef to infestation, so other species die too. Beyond the technical, data-driven loss to reef ecological balance is the wholesale theft of a reef's soul. Fish cue up, waiting their turn to be cleaned, to be as preened and primped as a prima donna by dazzling wrasses commonly called charismatic for their engaging personalities, their smiles, and friendly disposition.

THE CONSERVATION CASE ALSO HIGHLIGHTS the peril imposed on Hawaii reefs by removal of its yellow tangs—herbivores diligently grazing algae dawn to dusk to protect reefs from suffocation. 60% of the aquarium catch are yellow tangs—thousands daily—so that many reefs in Hawaii are now 100% yellow-tang-free zones. An irate collector in Florida called me, Snorkel Bob, to say, "You know and I know they're the cockroaches of the reef." I, Snorkel Bob, doubt that he knows much about life or reefs or yellow tangs.

THE CONSERVATION CASE ALSO HIGHLIGHTS the cruel reality that Moorish idols and many butterflyfish starve to death in 30 days with no sponges to graze, but they ship out daily with a 15-DAY LIVE GUARANTEE! AM I, SNORKEL F. BOB, WRONG TO CALL IT A MORAL ISSUE?

Note: butterflyfish mate for life.

Oval butterfly

Milletseed butterfly

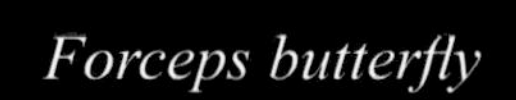

Forceps butterfly

Ornate butterfly baby

Ornate butterfly adult

Raccoon butterflyfish

Multiband butterfly

Teardrop butterfly

Moorish Idol

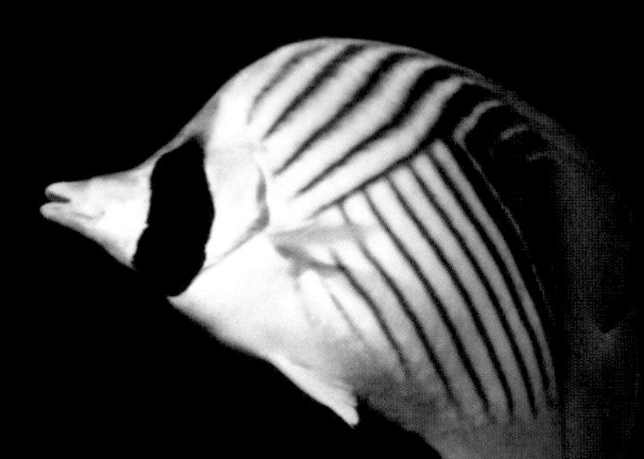

Threadfin butterfly

Pyramid butterfly

Can you guess which of the 3 cases against aquarium collecting prevails with most audiences? It's **THE ECONOMIC CASE**. Hey! I'm not complaining. I presented the sumbitch—but you have to wonder just how many good Christian values have gone by the wayside in want of economic justification.

Distant second was **THE CULTURAL CASE**, and I. Am. Elated! Why? Because people who think the Hawaiian culture was largely destroyed by resource extraction and disease can see history repeating itself on Hawaii reefs—even as the villains are swearing up and down that they are conservationists and this is GOOD for the reefs. I won't drag in another tired analogy from storm trooper times, but that *is* a bit like Adolph Eichmann showing up at cousin Bernie's Bar Mitzvah with a nice noodle koogle—chutzpa is a good thing and this ain't that. This is criminal.

Bringing us, lo and behold, to the smallest audience yet—those who appreciated the case against aquarium colleting based on conservation most of all. How can that be? Because conservation is a non-starter in the Political Will arena without economics and culture, because taking care of things must be *practical*! And if something cannot immediately converted to money or votes (power) you will gain NOTHING but lip service. *(We wish you the very best in your endeavors.)*

An important note here is that the Hawaiian culture case depends on the concept of *pono*, doing the right thing, which might draw an argument in a heartbeat, but deep down inside everyone knows what it is.

The Kona Coast is still called the Gold Coast, but now it's because of shoreline lots sold for million$.* It used to be for a priceless golden sheen in the breakers—tangs, that is, yellow tangs. Going, going, gone to the aquarium trade, where 99% die within a year of capture. They live to 40 years on a reef, keeping algae in check.

**Michael Dell (computers) built a house on the Kona Coast with a wall-to-wall aquarium. Do you really want to buy a Dell computer?*

... I think I know what's right. And wrong...

We all know what's right and wrong, yet some of us still struggle. In any event we *hele* on (heh•leh—to go, to move), lest we mire in dark thoughts when we could cruise with an abundance of fishes, including some old familiars, like Norton & Ralph—Ed Norton & Ralph Cramden, that is. Okay, a hint for you young'uns:

Hey, Ralphie, baby, have I got a deal for you!

Ed Norton & Ralph Cramden

I, Snorkel Bob, cringe at our culture's weakness for celebrity, especially in regard to those persons apparently famous for being famous. I will NOT mention names but will emphasize the vacuum of talent or intellectual acuity. Okay; Sarah Palen. But that's it! No more. What? Okay, I meant Sarah Palin. What's the diff?

Ed Norton & Ralph Cramden are an exception, because those guys captured a slice of Americana with verve, gusto, and formidable entertainment value. And so does Bart Simpson, another exception…

Staghorn damsel—Don't have a cow.

White-belly damsel—Geez, Homer!

Orbicular damsel—Kawabunga!

Orange-lined triggerfish

Speaking of old familiars, I just knew I'd met this guy before. And then it hit me: the ex-brother-in-law.

Hey, hotshot! You want in? Or you gonna play it safe again?

Which turned thoughts—even at depth on the Great Reef—to Wife One, whom I will absolutely positively NOT asperse here and now, though a question surfaced in a bubble or 2:

Why didn't she show me this side of things when we were dating?

Not really. I made that up—many people still think she's a lovely girl, and Ms. Cassowary didn't come along for another few days. Here's a fun fact: with saltwater crocodiles and so many deadly snakes in Australia, t-h-e most dangerous animal is the cassowary, given the hen's propensity to disembowel humans by ripping her spur claws up the mid-section. Who'd a thunk, with such a sweet puss and demure lashes? This hen was behind a fence, and I, Snorkel Bob, thought her a soft, sensitive lass. I've been wrong before, but I guarantee that this "deadliest/ most dangerous" folderol is trumped up by the media.

Think it over: would you rather snorkel with a salty? Snuggle a Taipan or a death adder? Or cross paths with a cassowary? Sure, a cassowary might kill more humans—because those humans don't sense danger in a giant chicken protecting her brood.

I, Snorkel Bob, heard Aussie watermen speak casually of white sharks and tigers, then go skittish recalling crocs in the surf. One intrepid fellow said they're as predictable as sharks but with only one behavior: kill. He said he likes to go camping on his days off, hiking up the hill with his girlfriend for solitude, relaxation, romance, and pastoral beauty. Trouble is, the campsite is a 4-hour hike, and the pesky Taipan snakes—small, abundant, lethal—allow a 2-hour lag on death following a bite. No worries, mate; you make sure your cell phone is charged up, and if you get nailed while gathering kindling, you call for an airlift. The chopper guys are great, hardly ever more than 45 minutes out.

I, Snorkel Bob, will give a wary cassowary a wide berth any day. What's she gonna do, chase me up a tree?

Meanwhile, back on the great reef, we may meet some fish who seem less than gregarious or even social—no small wonder, given the reef community's history of getting nailed for curiosity and acceptance of outsiders with apparent good intention.

It’s been a bad deal for the fish more often than not, with trust betrayed by a capture net or a hook, so a few sourpusses are only natural. Take this yellowfin damsel pup (left).

He’ll appear equally dour as an adult (below).

Which brings us to another milestone in our view of the fishes…

The media.

Many fans worldwide sense perfection in my, Snorkel Bob's, daily regimen. Go snorkeling, raise hell, yell at people on the phone, go snorkeling, throw wads o' dough at camera gear and exotic travel, go snorkeling, play with my pictures. It's not all fun and games, being the Greatest Living Snorkeler West of the Fertile Crescent, but greatness is never a sleigh ride.

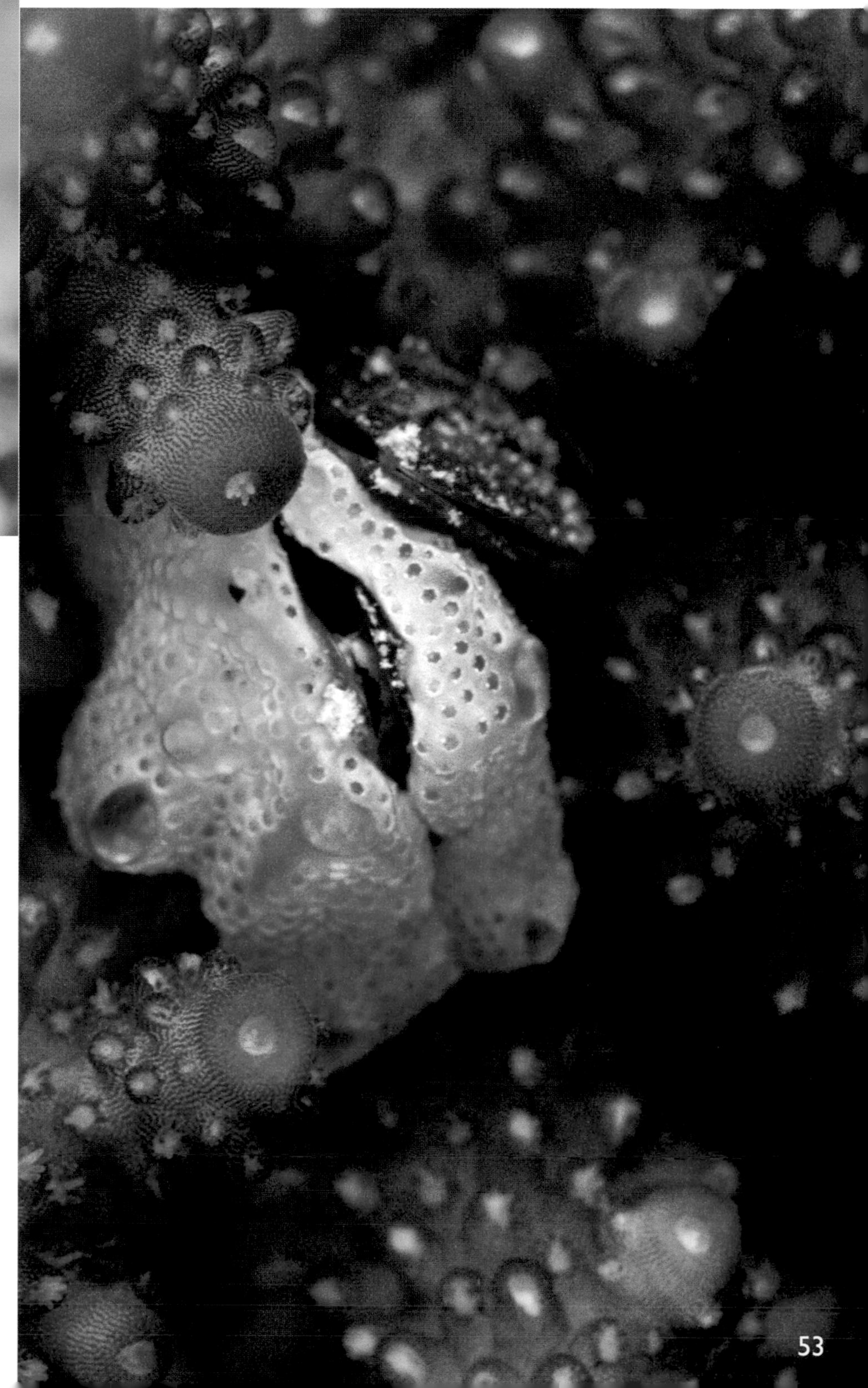

BE THAT AS IT MAY, many people on Planet E consider being Snorkel Bob as the most perfect job and fantasy fulfillment anywhere—not even Hugh Hefner sees more different shapes, sizes and colors.

I wouldn't touch this one with your keyboard.

Check the lips on this 2-spined angel—you think she's hitting the Botox? I think she's a natural. Note the similarity to Potter's angel, but down under she's a 2-spined.

Anyway, where was I? Where am I, Snorkel Bob?

Oh, yeah, pondering the reef, the fish, and the media—or show biz, such as it is. I, Snorkel Bob, recall the phone call not too long ago from a producer for the Discovery Channel. "I'm calling about an ocean adventure with Snorkel Bob, with unique perspective and voice."

"Indubitably. Why not?"

"All right then, what shall we do?"

"I don't know. I'll have to think about it."

"Good, but I can tell you, the executive producers (the money guys who frame ad sales based on program content) want danger. What have you got that's dangerous?"

"Hmm. That's tough. Truth be told, it's fairly tame around here (Hawaii)."

"What about sharks? You have sharks, don't you?"

"Sure, we have sharks. But I've never seen a tiger shark while snorkeling, and you can't very well call our reef sharks dangerous."

"So you do have sharks readily available?"

"Whitetip reef sharks, mostly, but they're pups. Sometimes curious but not aggressive."

Whitetip reef shark napping—after milk and cookies.

"Let's think about this," the caller murmured.

"Yeah, that would be best," I, Snorkel Bob concurred, briefly, till it hit. "Hey! I got it! The scariest encounter Hawaii has to offer!"

"What! Tell me! What!"

"Turds!"

"What?"

"Floaters. Blind mullet. Turds. Sewage. The State could ban charter boat dumping, but it hasn't happened."

(Linda Lingle administration [R], was neither kind nor gentle with reefs.)

"Do you want to get serious here?"

"I'm dead serious. I've cruised with sharks plenty. It's a rush—it feels like an honor. But when I looked up, eye-to-eye with a log jammer I nearly walked on water. I'd keep company with sharks over turds any day. Do you know how dangerous they are?"

"I'm not having this conversation."

So it ended as it had to end. How could the Discovery Channel sell a program based on sewage treatment failures and charter boat dumping in the tropics? It would be a stretch, but they did produce a vile series entitled *Vicious Killers of the Deep*, or *Deadly Ocean Killers*, or some silly, destructive title like that. One segment highlighted Hawaii's white tip reef sharks as menacing marauders out to eat you. Kee-riste on a crutch, that's like calling Lassie a jungle carnivore.

But here's the real rub: another segment featured "deadly" stingrays. WTF?

Well, we can call the executive producers at the Discovery Channel irresponsible, but that's too easy. Visiting the Great Barrier Reef, especially out of Port Douglas, you may hear several versions of the day the crocodile man, Steve Irwin, died. The story most credible to me came from a charter boat captain who'd worked that day, who said the production crew was out in a small boat (off a larger vessel) with Steve, looking for big animals, so

Steve could jump in and get filmed "interacting."

I don't know if Steve Irwin ever had a soft touch, though I doubt it, and he seemed compelled to dominate in most contact with wild animals. Granted, many were venomous or huge, requiring control. But in other cases, the control seemed gratuitous, inner driven. At any rate, he was a conservation champion despite any critique or quirk.

The charter boat captain recalled the VHF crackling that day with tentative sightings, confirmation, and chase. At one point a boat radioed to the production boat that a large tiger shark was seen a few minutes away, with GPS coordinates following. So the production boat raced over, and the camera crew slipped over the side with their gear. Directly over the site where the tiger had been seen, Steve jumped in. Steve Irwin had as little fear as any human could.

Most of the time the water on the Great Reef is clear enough to see the bottom—to see how the coral bommies rise from depth to very shallow. Apparently, the depth was less than anticipated, and Steve jumped right above a big ray, who instinctively raised his tail in defense. The barb at the base of the tail pierced Steve's chest—and heart.

Another version claimed that Steve Irwin approached the big ray quickly from the rear in an attempt to grasp the leading edge of the ray and go for a ride—what great footage! Except that it would be animal abuse, foolish and dangerous—and fatal. Fortunately, that version is speculative. Yes, it fits the characters of the human and the animal. The karmic point here is that the Discovery Channel segment on vicious ocean killers ended that day on a loss.

I, Snorkel Bob'm old enough to remember the Kingston Trio (barely), who first asked, *When will they ever learn? When will they learn?*

Gratuitously brutal programming on Discovery includes the "extreme angler" (Bah!) who leaps into murky water to torment big catfish and other toothy creatures for some mud wrestling and lip curling in a troubling display of macho dominance. In one morbid scene, the big fish lay gasping in a net while the extremist speaks of danger, teeth, death, ghastly potential, and on and on till you wish he'd vaporize. Finally, after compromising the fish to underscore the extremist's achievement and know-how, the extremist rolls the limp fish back into the water. This crap makes professional wrestling look nuanced and artistic. What is it with these guys? Oh, yeah: ratings and ad sales.

On the bright side, Jeff Corwin makes a point of respecting wildlife without the compulsory physical contact and macho control. What's not to like about that?

Meanwhile, back on the Great Reef, we move along to keep up, so we don't linger on a compelling mug and look up to wonder where everybody went and where we might be. Not lingering is difficult in most cases, since any reef addict knows that the fishes will take cover in the nooks, crannies, cracks and crevasses a reef provides when any big animal approaches. If the big animal hangs out, however, the fishes will peek out, then come out to see who and what.

The Great Reef is no exception but is so teeming with critters that moving along feels like a receiving line at a grandiose reef reception—every garish cousin and aunt and uncle comes forward to size you up and see what you brought. The cavalcade of fishes favors the coral for its cover, its backdrop and camouflage, and grazing potential. A typical pass through a coral bommie neighborhood might include...

Humbug dascyllus in antler coral.

Jewel damsel

emon damsel and soft coral

Bluespot damsel

Black-vent damsel

Blue Chromis (above)

Triangular butterfly (below)

Rainford's butterfly

A vagabond butterfly looks like a threadfin from Hawaii, but she has no thread fin trailing from her dorsal fin, and look at her tail: yellow and black. She has subtle iridescent piping in the latter half of her dorsal fin, and likely a few other quirks distinguishing her from a threadfin.

Redbreasted Maori wrasse

Cryptic wrasse juvie, I think

Slender wrasse, maybe

Okay, left is a cryptic wrasse, I think, unless that's a dimple on his hiney instead of the telltale cryptic freckle. And above is a slender wrasse, maybe. If not, he's at least a skinny wrasse. At any rate he's a very small reddish fish who agreed to pose for a reasonably good shot. Maybe he figured stillness was his only hope as I honed in for the close-up. What's notable here is that I, Snorkel Bob, went for years without knowing the difference between a wrasse and batfish—or a rat's assfish. Okay, that's an exaggeration, and I am trying to cut back, but I didn't know the fish like I do now and it was, is, and shall be A-OK. It's good to know the fish, but you're always going to meet a fish you have to look up anyway. The key point is that it's the behavior, not the species or Latin binomial that grants insight to Neptune's ways and the mysteries of the deep.

You'll note that most fish ID books only dabble in behaviors. That's good, because two individuals of the same species can be shy and gregarious—scared snitless or your new best friend. As in plants and people, personalities tend to run the gamut in fishes too. Like people of all stripes, no fish chooses containment in a glass box—that fate requires nets and greed.

Below are bridled monocle bream, abundant and curious, hardly engaging—but a sociable monocle must be cruising out there somewhere.

adult

juvie

Like angelfish everywhere, the Great Reef angels stand out with spectacular colors and patterns. Like angelfish everywhere, they are oppressed by aquarium collectors filling orders for exporters, resellers, retailers, and home hobbyists, some of whom don't give a horse's patooti for me, Snorkel Bob, or anything I have to say about voiding reefs or stealing souls or moral issues. Then again, some do.

Bi-color angel (above)

Emperor angel (below)

6-banded angel

You can see why juvenile juicy tidbits in boysenberry, lemon, and butterscotch…

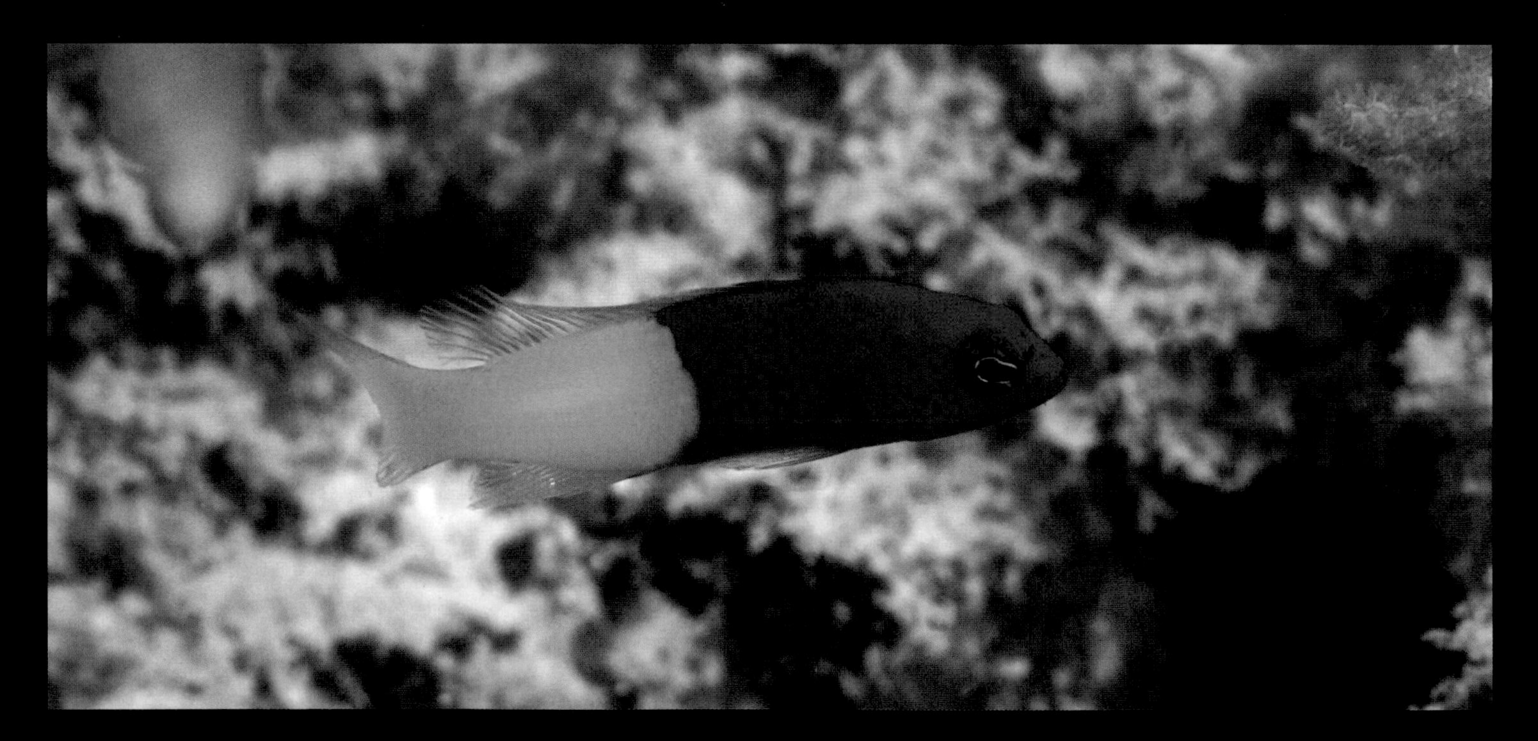

Royal dottyback & Zebra dartfish

Baby perch

…wouldn't want to get caught out.

Emperor

Red emperor

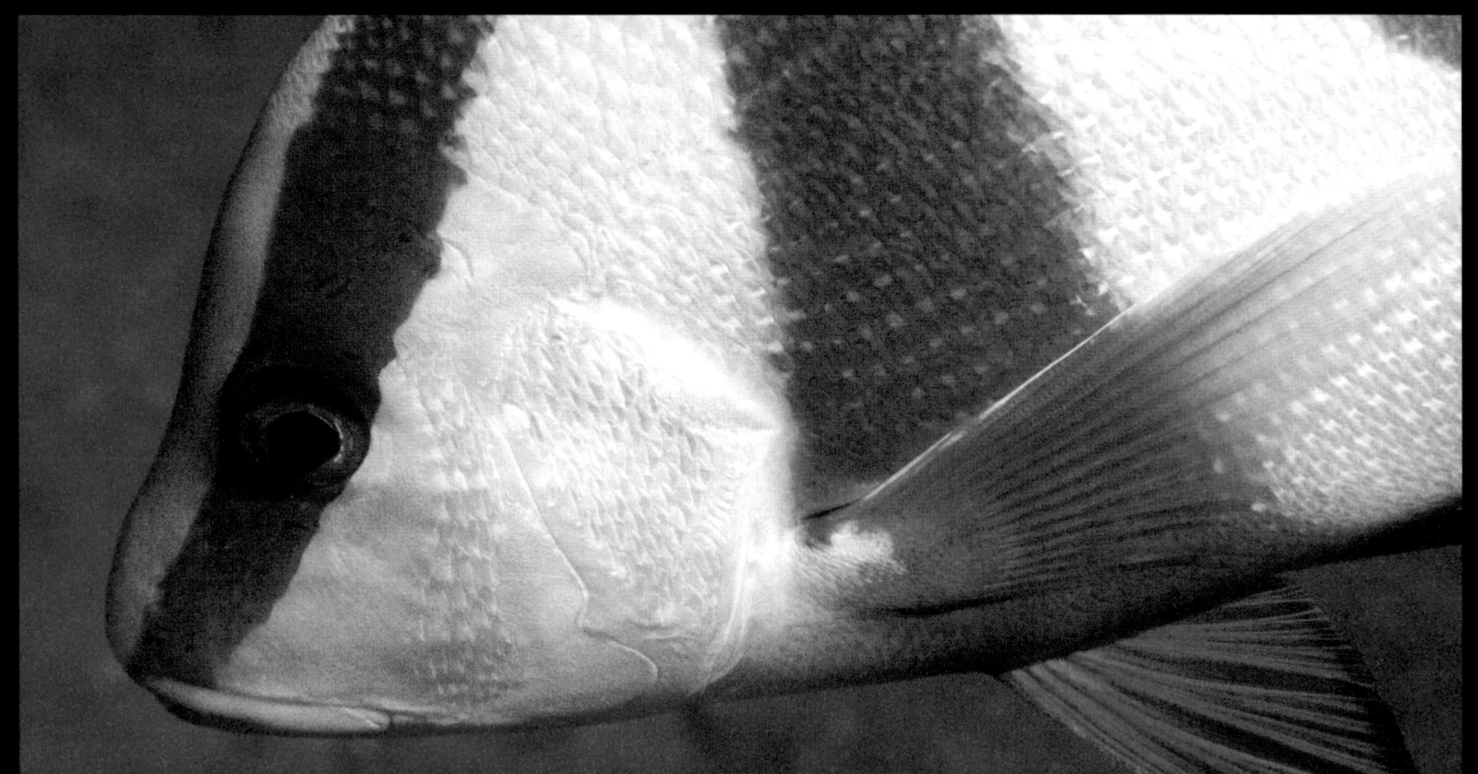

Harlequin tuskfish

Barred rabbitfish

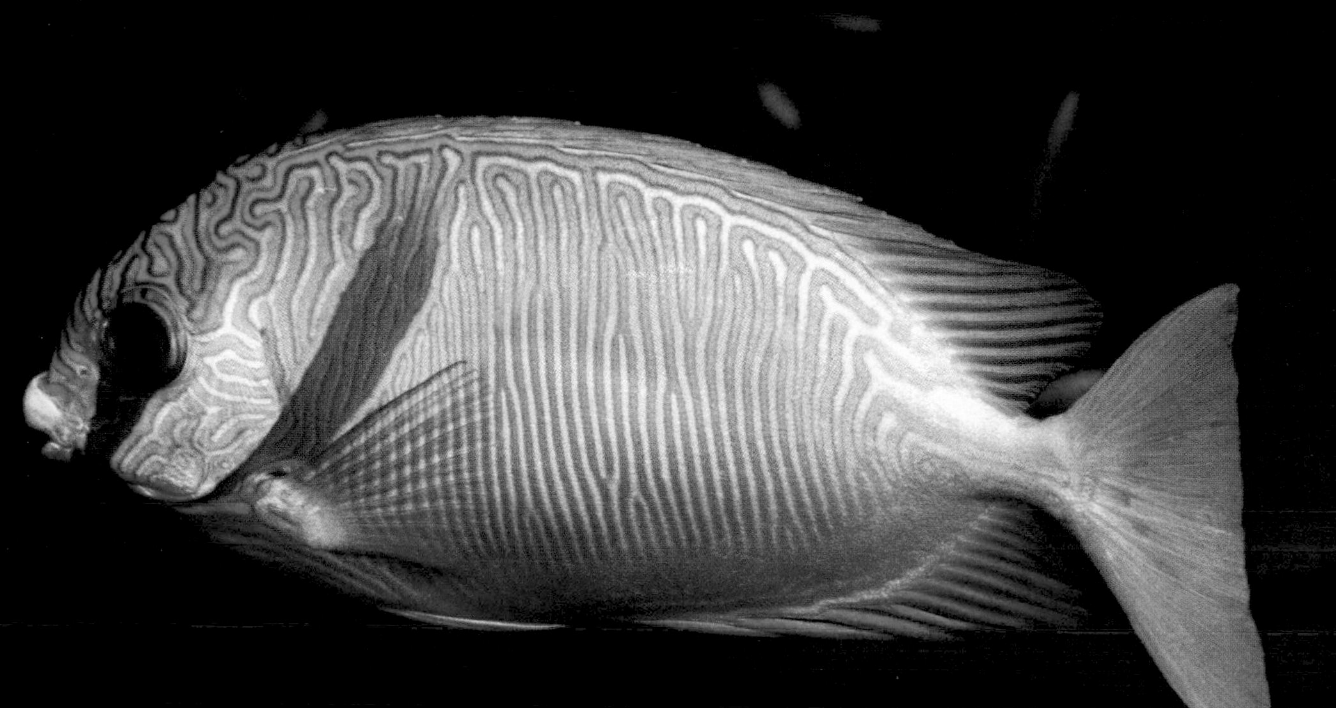

Foxface (above)

Yellowtail—schooling game fish, on the hunt. (below)

Bottom dwellers favor stillness to render complaisance in unsuspecting prey.

Longfin rockcod—Hello, little fellow. Are you lost?

Speckled sandperch—Maybe I can help.

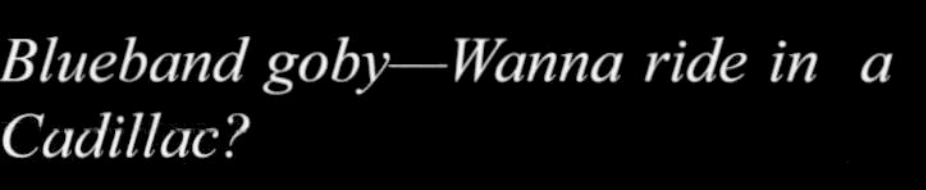

Blueband goby—Wanna ride in a Cadillac?

Cleaner fish keep a reef free of parasite infestation by picking tiny bloodsuckers from the skin and scales of bigger fish. Wrasses do the heavy lifting on some reefs, while blennies prevail on others—and herbivores pick mossy goobers too. This unique cleaning behavior benefits reef society in general and parallels land-based society by spawning charlatans—scammers who would have you believe, only to take advantage. *Do you own your home? Do you have a credit card? Do you like vacations? Do you need vacations? Do you want vacations?* Like that.

Here are two reef poseurs making a go on the Great Reef:

Bluestriped fang blenny—Yeah, one owner, a Sunday school teacher, that was it, a teetotaler who mostly stayed home to watch PBS. And meditate!

Mimic blenny—I'm from Nigeria and want you to share the $3.9 million my uncle left me!

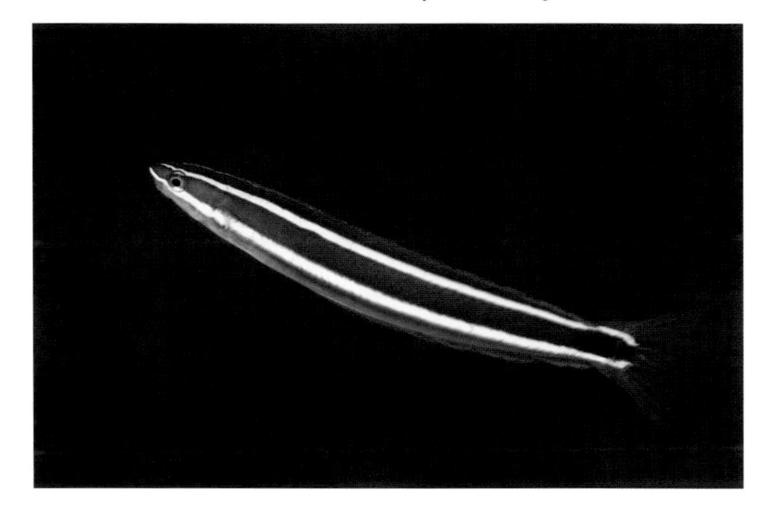

These guys set up shop with body language or chemistry or who-knows-what mysteries of the deep to attract fish for a cleaning—and deliver a reaming. Once in place, the mimic or fang blenny will open wide on formidable fangs to chomp a gob of snot, scales, skin, goobers—whadda ya got? Yummeh! Then they run for the hills, because, cute as they are, you want to wring their scrawny necks. Oh, yes, they will move on a snorkeler or diver too.

Is there a timedog blenny with gold nuggets on his rings and watches, waiting to be discovered on a far yonder reef? *You can trade this Rolodex card for a week on the top floor anywhere in the world!*

Black-spotted puffer

The Great Reef is a sign of the times, a last vestige of nature in effusions more prolific than most humans have ever seen. The Great Reef is a last hope for survival in our time.

In *Ghost Dog* Forrest Whitaker plays a hit man who lives by the bushido code as written in the *Hagakure, the Book of the Samurai*. It's better than most movies and stuck with me, Snorkel Bob. We can boycott Asian products, ban plastics, and recycle every day, but nature keeps dying. We go beyond the spirit of any age—burgeoning humanity is killing the Great Spirit.

I do think every aquarium should come down, because every fish on every reef is a soldier in the struggle for recovery.

As reefs decline, however, I often think of the Ghost Dog and the *Hagakure*:

It is said that what is called "the spirit of an age" is something to which one cannot return. That this spirit gradually dissipates is due to the world's coming to an end. For this reason, although one would like to change today's world back to the spirit of 100 years or more ago, it cannot be done. Thus it is important to make the best out of every generation.

Soft coral

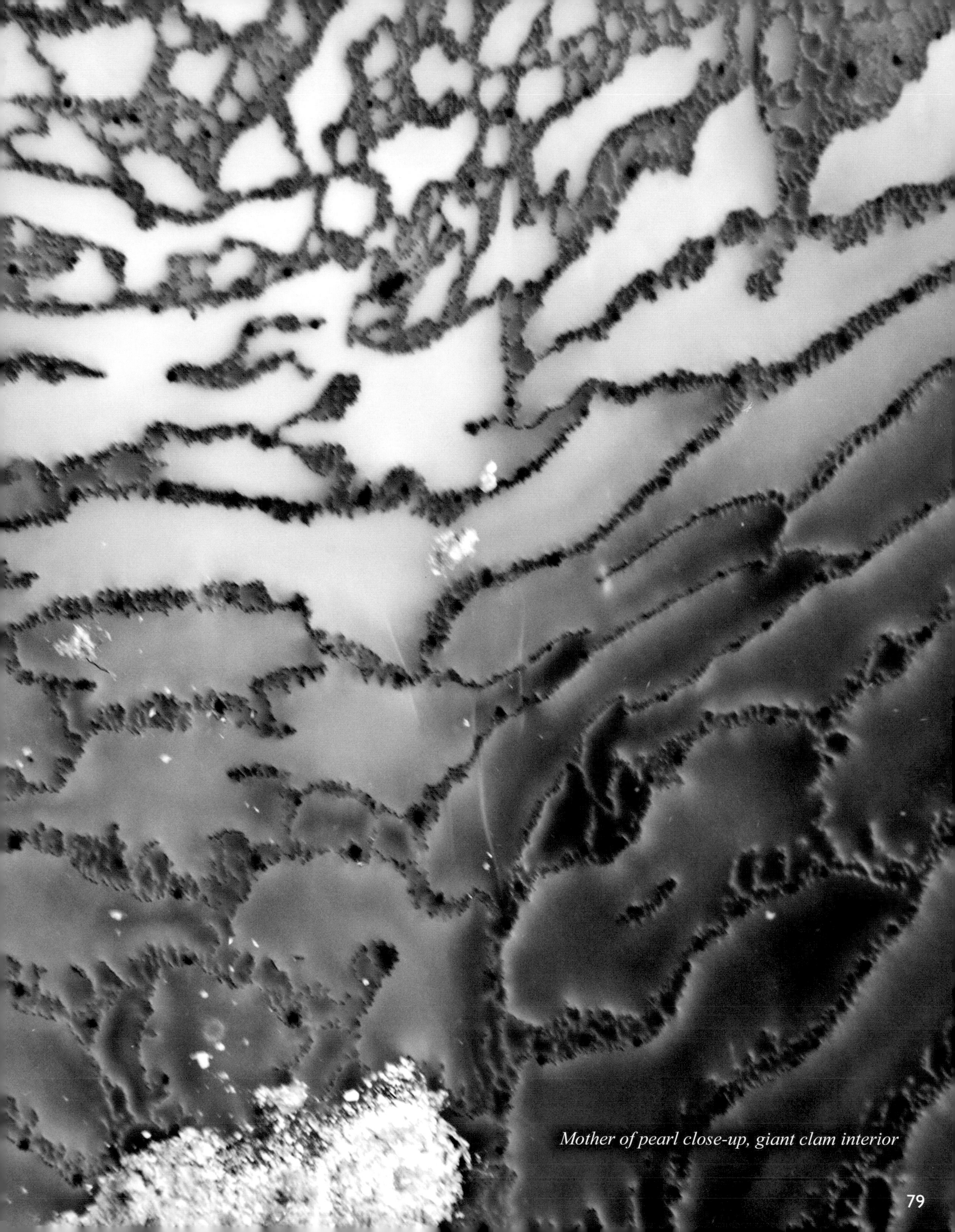

Mother of pearl close-up, giant clam interior

Soft coral

On a brighter, more uplifting note, I, Snorkel Bob, would like to share a tale of joy and hope. Here too we begin with distress but end with optimism.

I.e., not so long ago I, Snorkel Bob, got a gallstone stuck in the bile duct—a big, greasy sumbitch throbbing with pain like a coronary thrombosis. Twice in three nights sweet slumber slammed into pitching sweats and angina. No, kids, that's not where babies come from—it's what can take Daddy away. Angina is chest stricture, often caused by cramping heart muscles.

The gallbladder resides near the heart; hence the similar symptoms. But heart attack usually comes with shooting pain down the arms, across the shoulders. I had no shooting pains and besides: a regular snorkeler with a weak heart? I doubt it!

Long story short, laparoscopic removal had me, Snorkel Bob, back home in five hours flat, in time for Oprah. But all was not well.

The offending gizzard sat forlornly on a shelf till I, Snorkel Bob, felt pangs of sympathy for Gallbladder, all purple in the gills, staring sadly from his jar, while I had an adventure to plan.

Then came inspiration—a transplant in the spirits of reef recovery and loyalty between old familiars. After years together of thick and thin, lean and fat, Gallbladder and I, Snorkel Bob, would keep the bond. Hey, inflammation can happen to anyone.

The arduous flight to Oz included a layover in Guam. Gallbladder flew coach, more cramped than his jar—TSA would not allow the jar but required a boda bag. The ignominy!

Okay already, I made that up, but the figmentation materialized in a blink on first glimpsing these ascidians at home on the Great Reef. Can you blame me, Snorkel Bob, for the instant association and Spirit of Reef Recovery?

Will you look at those throbbing purple veins on the slightly jaundiced backdrop? *Feel* the pulse of seawater and bile pulsing in and out—*sense* the bulging indication of recalcitrant cholesterol globules, each with a mind of her own and a keen eye on revenge. *See* the surgical precision bearing down on the troubled organ, removal of which is not so simple: "If we get in there and discover that the thing is attached to the liver, then we'll need the full abdominal incision to give us more room to get the thing off." THIS DISCLOSURE came as the gurney rolled into the O.R., cold as a walk-in freezer (to discourage antibiotic-resistant bacteria that can kill you) and more harshly lit than FOODLAND at midnight. They spared the patient this harsh view of the O.R. until recently but now aim to minimize inadvertent injury in transferring

Transplanted at last and throbbing where he won't cause any pesky chest constrictions—Gallbladder adapts.

What ho? It's the mating stance! It might as well be spring down under!

Mr. & Mrs. Gall F. Bladder & children at home.

a generally anesthetized body from the gurney to the operating table. So they have the patient transfer himself while conscious. The surgical platform is also cold, and narrow enough to fit between the shoulder blades, which provides a comforting distraction, allowing the patient to think of other things, like rolling off, instead of the ghastly exploration dead ahead.

The surgeon's disclosures on liverish complications are followed good-naturedly by the ri*dic*ulously low odds on occurrence, along with DEATH, MORBIDITY, and ETC.—"At least it's never happened to me," me in this case referring to the surgeon's patients over the years.

Then a fellow as yet unmet leaned over to murmur upside down, "Here's the good part…"

You wake up feeling shot at and missed and shot at and hit—I mean really, like a kick to the solar plexus with a steel rod jammed down your throat. Okay, so it was a tube, not a rod, and it was plastic, not steel. Intubation was not disclosed, and I will not belabor this invasive, over-the-top exercise in covering THEIR ass at your most personal expense here, except to say that I will not allow it again.

And yes, Gallbladder was saved in a jar and could have made the flight but *come on*. Can you please get a grip on reality?

These things are ascidians, also known as tunicates because of their tough, leathery skin or tunic. They're also called sea squirts, because they squirt when stepped on, perhaps by Asian restaurateurs. Most amazing, according to the Australian Museum in Sydney, is that ascidians "are not invertebrates, but primitive chordates, the phylum to which all back-boned animals belong. This evolutionary relationship to back-boned animals can only be seen in their small short-lived tadpole-like larvae, which swim for just a few hours before settling to grow into an adult firmly attached to a rock or weed." What is the meaning of these boneless branchial baskets of brine belonging to a vertebrate phylum? Beats me, Snorkel Bob, unless maybe it takes some backbone to bear up to such a bald-faced resemblance to a gallbladder.

BUT ENOUGH OF SPINAL SPECULATION!

We're off to…

Common to these reefs, the longspine squirrelfish seems tentative but is approachable and can't help the sailfin flourish. What a show!

ST. CROIX, U.S. VIRGIN ISLANDS

Where the biggest tourist magazine informs us:

"MARINE LIFE: All native plants and animals of the USVI are protected under the Indigenous Species Act. It is illegal to take, catch, possess, injure, harass, or kill any native animal or plant. All fish (unless harvested for food), coral, shells, and other marine life are protected. You will not be permitted to leave the territory with shells, corals, or any other marine life in your possession."

Soon after Hurricane Hugo (1994) you could drive along Cane Bay where many shoreline dwellings were abandoned, with nothing left but the ruins. You get the same view today. Calling St. Croix depressed, however, is only accurate for the dilettante elite of the Chamber of Commerce.

With 3 cruise ships per week instead of several per day, like St. Thomas, you'll see far fewer fat people on St. Croix waddling here and there, stretching things out between bouts of two-fisted eating. And since Frederiksted, where the cruise ships dock, shuts down between ships, the place has an eerily soulful ring to it.

What's that sound? Oh, yeah, it's no people choking the sidewalks, looking for chachkas. The sound goes away when the ship is in. But, thank Neptune, the ship leaves at 5, once all the fatties have waddled safely back aboard to prep for some snax before cocktails and din din.

Okay, I, Snorkel Bob'm no fattist and could stand to lose a few lbs myself—maybe 1½ or 2 anyway. I only mention the cruise ships and fat people because let's face it; those blubber barges would shut down if Baby Huey couldn't belly up to the buffet for breakfast, brunch, lunch, dinner and late snax—oo!—how 'bout a baloney sandwich at midnight, for the sinful, delicious taste of the thing.

photo by Andi Christie

Hey. Why not at 10? And again at 11! With cheese! And bon bons!

But here's the kick: you look at one of these sideways condominium towers and can't help ponder its gargantuan role in the obesity epidemic in America as well as a sewage disposal problem of formidable magnitude. The levels of consumption and waste are so HUUUUGE that you may forget to wonder what lies just below the surface.

You can't actually go into the water for a look when the ship is in because of security, though I, Snorkel Bob, wonder who in the world would dislike the United States of America and want to harm a cruise ship fattening up the gentry. My GAWD that's harsh.

So we'll move on down to the nitty gritty with a tale to warm a snorkel curmudgeon's heart. The pier where the cruise ship parks is supported by concrete pylons, all coated with many different sponges, soft corals, hard corals and variable goo and schmutz of the deep—don't worry; no toilet paper. I, Snorkel Bob'm acutely sensitive to dirty reefs. (See Vicious Ocean Killer Teeth Death Danger Blood 'n Gore Murderers, page 60).

The pier is a most popular dive spot because of its concentrated habitat for small critters—and because of its navigational charm. Even I, Snorkel Bob, would be hard pressed to lose my way here. You follow the pylons out, or you follow the pylons back. Okay, I got confused once but applied my degree and worked it out.

But at 50' depth on the far end and 25' for most of the middle, the pier pylons got thrashed in Hugo ('94). What did the reef community do? A reef community is both above and below the water, both fishes and humans who live, love, work and protect the reef as their own—snorkelers, divers, pedestrians, whackos, droolers, scientists, fools, dreamers, tourists, even a Republican or 2, in spite of the naysay and rhetoric.

The reef community humans gathered the remnants of habitat—sponges, corals, sea fans, schmutz and goo torn from the pylons and scattered about on the bottom. Then they pieced these remnants together along the pylon sides and tied everything in place with monofilament line that dangles to this day. The habitat attached itself in time and filled in.

Never was a heart so warmed as when I saw this vibrant community of fishes restored to former splendor by vibrant human hands. It can happen. It happened at the pier in St. Croix.

Pardon me, Snorkel Bob, for the repetition, but this puffer—and all puffers—remind me of the testimony presented by an aquarium collector to the House Water & Land Committee of the Hawaii State Legislature during hearings to regulate aquarium extraction: "We see puffers on almost every reef. So why shouldn't we take them?" That was January '08. To this day Hawaii has no limit on aquarium catch, no limit on the number of catchers, and no constraint on endemic or rare species. Payoffs, corruption, backstabbing, dirty dealing, corkscrewing?

Wildlife trafficking for the pet trade is more rampant in Hawaii than in any other state in the U.S.A.

You might think these tube sponges need to see a urologist, but I, Snorkel Bob, assure you they're in the pink, and maybe just a tad yellow. Though few critters seem to live in these cavities, some critters use them to hole-up, as it were, on the approach of a bubble-blowing humanoid.

Just for fun, I, Snorkel Bob, did a colonoscopy on this tube sponge—not to worry, the opening was HUUUGE, though the terrain seemed inhospitable, clearly more suitable for brief refuge than permanent dwelling.

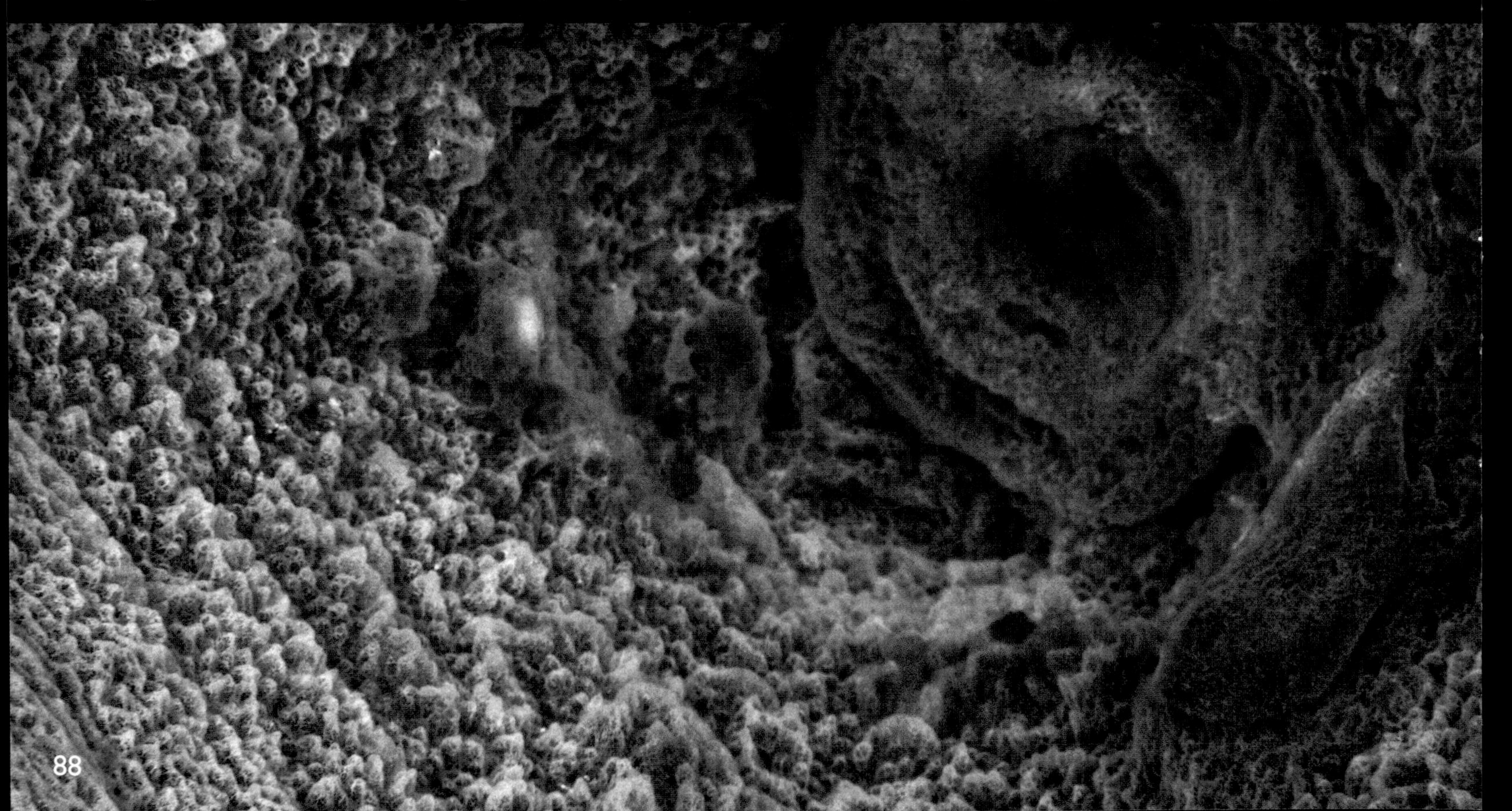

But most fish at the pier enjoy a sheltered life with few predators, kind of like an aquarium, except that they don't die in short order after a long flight and many rounds of rough housing that render reefs EMPTY! Don't get me started.

Keep me composed and happy in this fantasy realm, where small puffers swim freely in open water—a rare behavior in the wild blue yonder, beyond the pylons.

The carefree swimmers among the pylons are reassuring indeed. Forgive me, Snorkel Bob, for being a speciesist, for going gaga at the rare and exotic, when all fish are created equal and want nothing more than life, liberty, and the pursuit of happiness, but…

Gaga!...

Gee, Snorkel Bob, it's an excellent swimp on a hub, I mean swimp on hub, I mean shrimp, but gaga?

To which I, Snorkel Bob must remind you that these swimps breed like rats!

Wha? You think.

You still don't get it—you think I'm going gaga for some swimps, I mean shrimps, about as big as a gnat's patooty?

Wait a minute! Swimps as big as a gnat's patooty? That means…

Seahorses and all fish have no eyelids and can be acutely sensitive to shock and/or trauma. Sharks have protective eye covers that rise on feeding, but that's hardly the same as a flash on a seahorse. This seahorse didn't cringe, but it was three shots and out. NOTE the open pouch hole, indicating that this is a male. She deposits the eggs into the pouch. He hatches the babies. ALSO NOTE how the pouch hole resembles the hole in the red tube sponge—a favorite hitching post.

Note the brood pouch here too—another male, deeper and darker but swaying gently with the current just like his coz. Why do these fragile grubs seem so mystical?

Can it get any better? I don't think so, unless we happened to stumble onto…

Christmas trees are small polychaete worms who feed and breathe through spiral plumes of many brilliant colors that vanish in a blink on sensing a presence.

Sea anemone at the base of the Frederiksted Pier on St. Croix

A banded butterflyfish grazing amiably among the small fry.

A less amiable, schooling pelagic who likes to gang up, run fast and hit hard, may be cruising the microcosm among the pylons for the stray snack.

This porgy's appetite is only practical, even in such a feel-good place, but don't you worry, flashing strobes light this guy brighter than a neon sign: **EAT AT PORGY'S**.

And you never know who might be passing just yonder on the interstate feeling hungry…

Pardon me. Would you have any Grey Poupon?

Shemolies! At least this apex predator is well-mannered and approaching lunch with aplomb. As if that's not enough to scare the holy screaming Cheez-its out of a mild-mannered snorkel exec, check out this guy's bridgework. But what seems hazardous to humans is only casual on a reef. Hardly a flutter to the next pylon, we come to...

...a scorpionfish, apparently in a reasonable mood, though the dorsal spines on this little bruiser can cause severe pain and illness. These spines do not suffer erectile dysfunction and can be a hazard in the shallows.

Which might just wrap things up under the pier. Here's a junior razorfish, a shy, tentative fellow who peeks out every few seconds to see what's taking us so long to leave…

…so we'll head out, except for the squiggle at the base of that pylon in the stems and nubs just yonder…

Could it be?...

Oh, are you still here?

A spotted drum baby!

Spotted drum juvie

Jackknife juvie

It's no secret that I, Snorkel Bob'm an equal opportunity fish fancier. That is, I don't give a snit about species differentiation based on tiny iota. HOWEVER, to keep the more scientific among you happy, I'll point out that this 2nd juvie is not a spotted drum but a jackknife fish. What's the diff? The spotted drum juvie has a spot on his nose, while the jackknife fish juvie has a vertical line on her nose. So I'm told—though the difference to me is negligible. NOTE here that stripes are horizontal, while bars are vertical. The spotted drum is striped, while the jackknife is barred.

A jackknife juvie comes forth to pose this way and that, showing off her frilly fins and demure indifference.

Kids grow up, in this case with spots and more body bulk.

Okay, it can't get any better than that. Can it?

Smooth trunkfish

This is a face only a mother could love, and so could you and I, Snorkel Bob.

These two (above) are also *smooth trunkfish*.

Half a mile south of the pier are several healthy reefs where a *spotted trunkfish* or two graze…

…along with a gregarious cowfish—note the horns or spikes over the eyes.

This cowfish has more of a piggish puss along with her amoebic green skin and eyebrow horns. While that would make some people self-conscious and possibly reserved, Ms. Hambone here couldn't get enough of center stage. Well, it's not every day a break in showbiz comes along, and the audition queue fairly stacked up down the block and around the corner. She may be all puckered up for a smooch, but she's got some stiff competition…

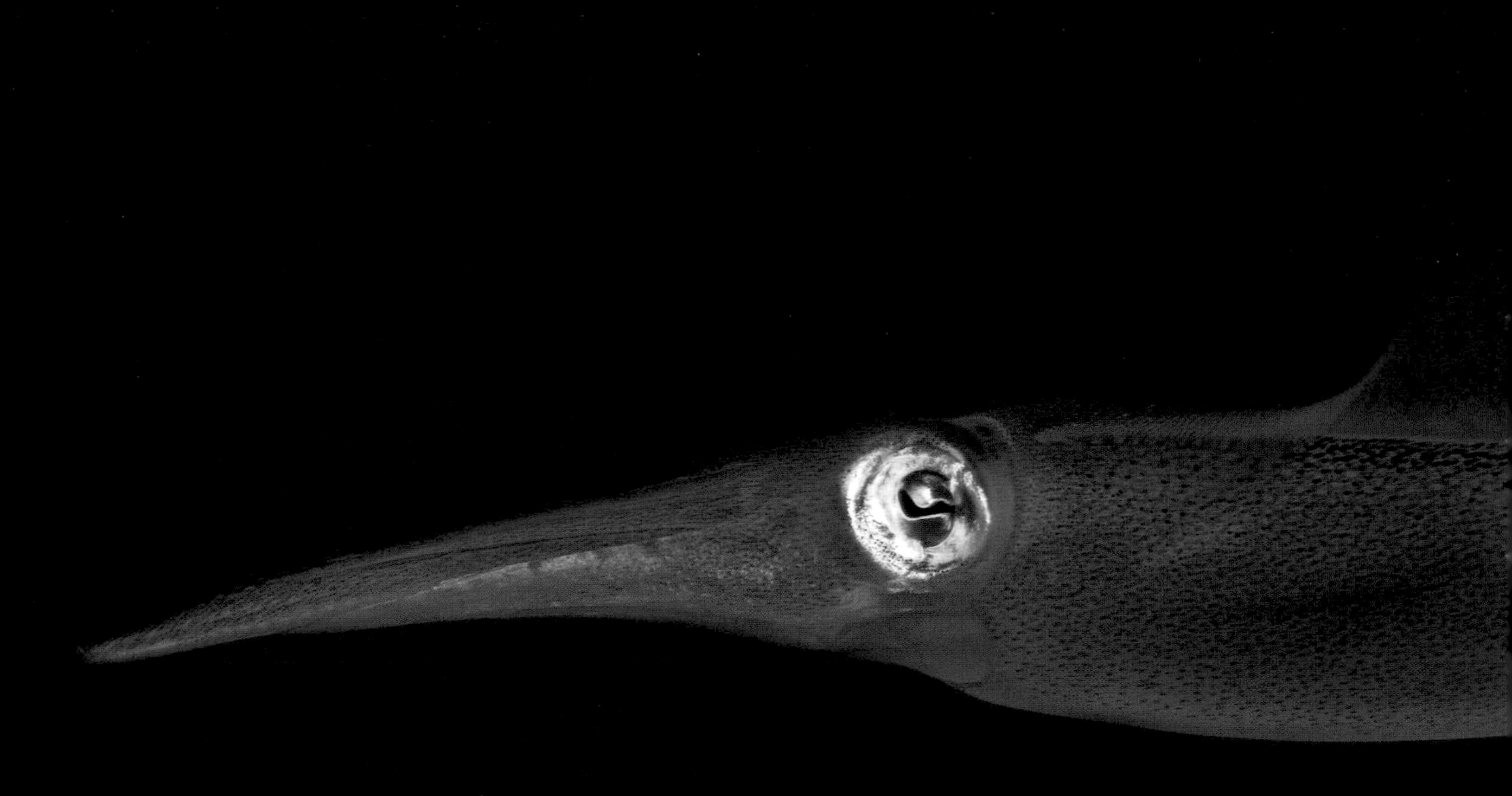

What a profile. The thing about squibs, I mean squids, is their ability to change color, pattern, and intensity. I've seen them go fuzzy green, like detritus flecks, then flash into mottled pink on blue—make that black, hazel, gray, salmon, red, brown, flesh tone. Flesh tone? Like Band-Aids for Caucasians? She looks good in the face, but will she hold up for the full figure?

Then again, on the downside of showbiz, you got your drug abuse. Fortunately, this little gill breather made it through rehab and now stands strong, even in the face of temptation.

He's a glass goby adult, like this school hovering near a coral head, darting for plankton.

Okay, so the whole showbiz-drug-abuse schtick was ill-advised, of questionable taste and/or appropriate content for a family audience, but you got to admit, this little gill breather is a ringer for Al Pacino in his classic movie *Bug Face*. No?

This is not Al Pacino but a coney (sea bass family) in the brown phase with isopods (parasites) attached.

And the glamour front heats up any time you get blue chromis.

These blue chromis seem more similar to Tahiti chromis than to Hawaii chromis in color—dazzling blue here—though the assertive disfavor is common to most chromis I've met.

At the apex of showbiz glitz is a fish we don't see in Hawaii anymore. The fire dartfish put on a great show in Hawaii too, till aquarium hunters got them. But I dasn't tarnish an intro with such color on hand… Ladies and gentlemen, boys and girls, *I give you fairy basslets…*

Finally, for you aging veterans of the Summer of Love, I did not do this intentionally. In fact, I don't remember taking this picture, PROVING I was there.

NOTE: Caribbean reefs appear to host far more sea bass than other places. Sea bass include many species and families, like grouper and basslets ranging in size from mere inches to mongo humongous, from the pygmy sea bass at 1½" to the Goliath grouper at 8'. The best known sea bass may be the Chilean sea bass, fished and eaten to near extinction and still on the brink. Chilean sea bass sells so well, however, that Whole Foods established its Marine Acceptability Council, or some such hokum, to "approve" continuing plunder. You'll also likely see swordfish in the Whole Foods seafood display, and you may have heard the owner/CEO of Whole Foods reveal his inclination on healthcare reform. His rant makes me, Snorkel Bob, sound like Miss Manners.

I, Snorkel Bob, think healthcare reform a good idea, and plundering Chilean Sea Bass is a crime against nature. Selling swordfish is irresponsible; it's full of mercury, and swordfish longlines are killing what few sea turtles remain. Here's hoping that you can step around these nasty droppings on your path to the grocery—and don't forget further color, erudition, and entertainment, like with this splendid little bass in Cane Bay…

Harlequin bass generally top out at 2½-3½", and like all bass, basslets and groupers, have long bodies and big bass lips.

Cane Bay is on St. Croix's north shore, and though snorkeling there is adequate, the perennial shore break keeps things stirred up most of the time at snorkel depth.

At 40-50', however, clarity prevails, along with flora and fauna indicating a healthy reef community.

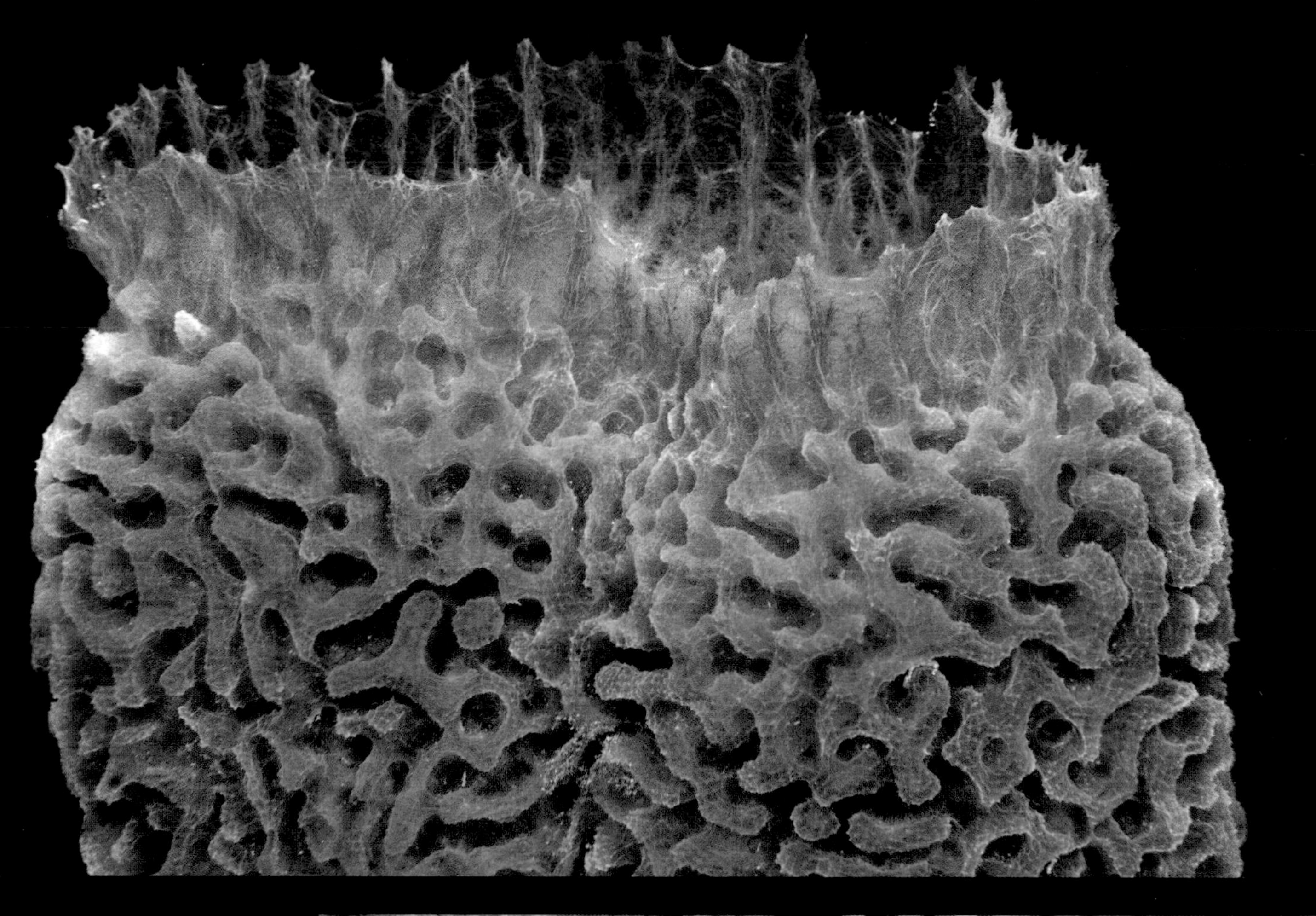

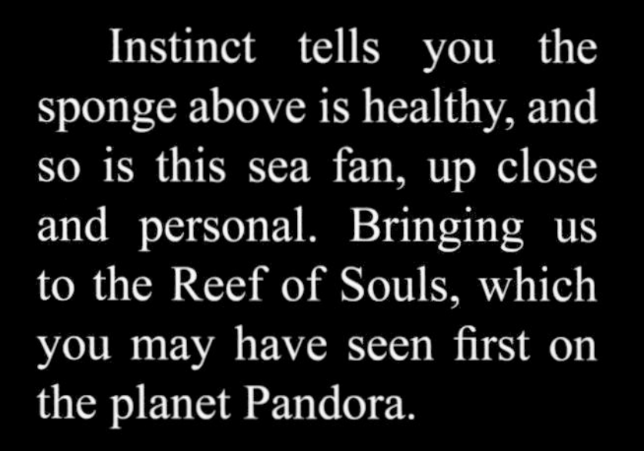

Instinct tells you the sponge above is healthy, and so is this sea fan, up close and personal. Bringing us to the Reef of Souls, which you may have seen first on the planet Pandora.

Commonly called a Christmas tree hydroid, this soulful being is slightly toxic—if you're a human, that is.

Vibrance abounds with species variation, color and balance. How many species co-exist in a 12” radius of this blenny’s home?

How many species live within a proportionate radius of your home? How many of those species show their health like this?

It's a reef caterpillar with fuzzy balls.

(What? You expect me, Snorkel Bob, to know everything?)

Okay—it's a fire worm and yes, the fuzzy balls are hot—stinging—to the touch.

Besides abundant indicators of health and well-being, the reefs at Cane Bay host some old familiars, like this slipper lobster, trying to fit in…

Caribbean slipper lobster…

And just for fun, here's a Hawaii slipper.

These St. Croix butterflies also look familiar:

Longsnout butterfly

4-eye butterflies

Two more old familiars are this *spotted moray* and *lizardfish*

Perhaps more unique to the Caribbean Sea are these garish pastels.

Spanish hogfish

Bluehead wrasse

These three wrasses are the same species in juvie, initial and terminal (full-grown) phases.

Yellowhead wrasse juvenile

Yellowhead wrasse intitial phase (below)

Yellowhead wrasse terminal phase (bottom)

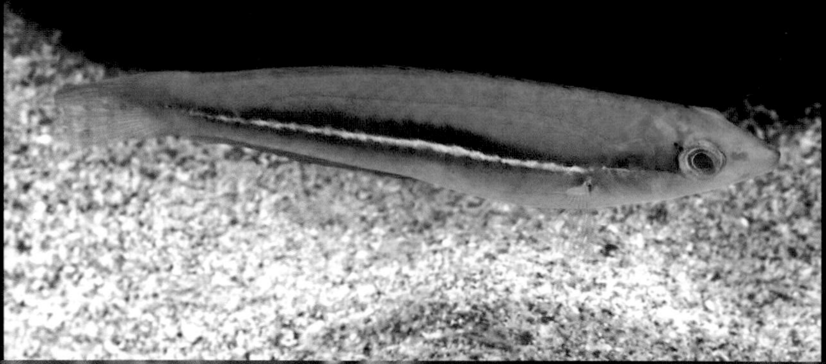

The wall at Cane Bay drops vertically from 95' where I, Snorkel Bob, got this portrait, up close and beautiful of a *rock beauty angelfish*.

photo by Anita

On Buck Island is more harsh encounter with the cold reality of Planet E. Global warming is not real if you have a glass navel.* The bleaching event of 2005 killed 30-35% of Caribbean coral that year—in some areas the kill went to 80%. Shallow water warms faster than deep, with less filtering of direct sunlight. Most reefs have little temperature tolerance range, so the Caribbean, also known as the shallow sea, is more susceptible.

* to see through, because your head is so far up your pooper chute.

Coral is comprised of colonies, live polyps growing on the skeletal remains of their forebears; the primary component of coral is calcium that weakens when the coral dies. With no life, the dead reef collapses. Most reef addicts have seen dead coral, but dead coral never looks so tragic as dead elkhorn coral, a most spectacularly intricate habitat.

It's slim pickin's all around, and a depressing indicator of what's ahead. Even this 7' lemon shark looks a little close to the bone and wouldn't turn around to smile for the camera.

Please don't get me, Snorkel Bob, started. Suffice to say at this juncture that 6,000 miles to the east of Hawaii in the shallow sea, a 1st-coz species to the yellow tang is the blue tang.

Note the juvenile blue tang is elusive, shy and as YELLOW as his cozzin way yonder:

Among indications that this reef is trying to recover is this cleaning station. Blue tangs and cleaner gobies are integral to reef recovery, reef health, and reef stability. I, Snorkel Bob, often hear parrotfish profiled as the lynchpin species of any reef.

Parrotfish maintain the reef by creating substrate receptive to new coral growth. They do so by grazing on the front end and by spewing sand out the back end. Cleaner fish—gobies in this case—maintain the maintenance crew. This is balance.

…and beauty.

Other notable reef denizens at Buck Island:

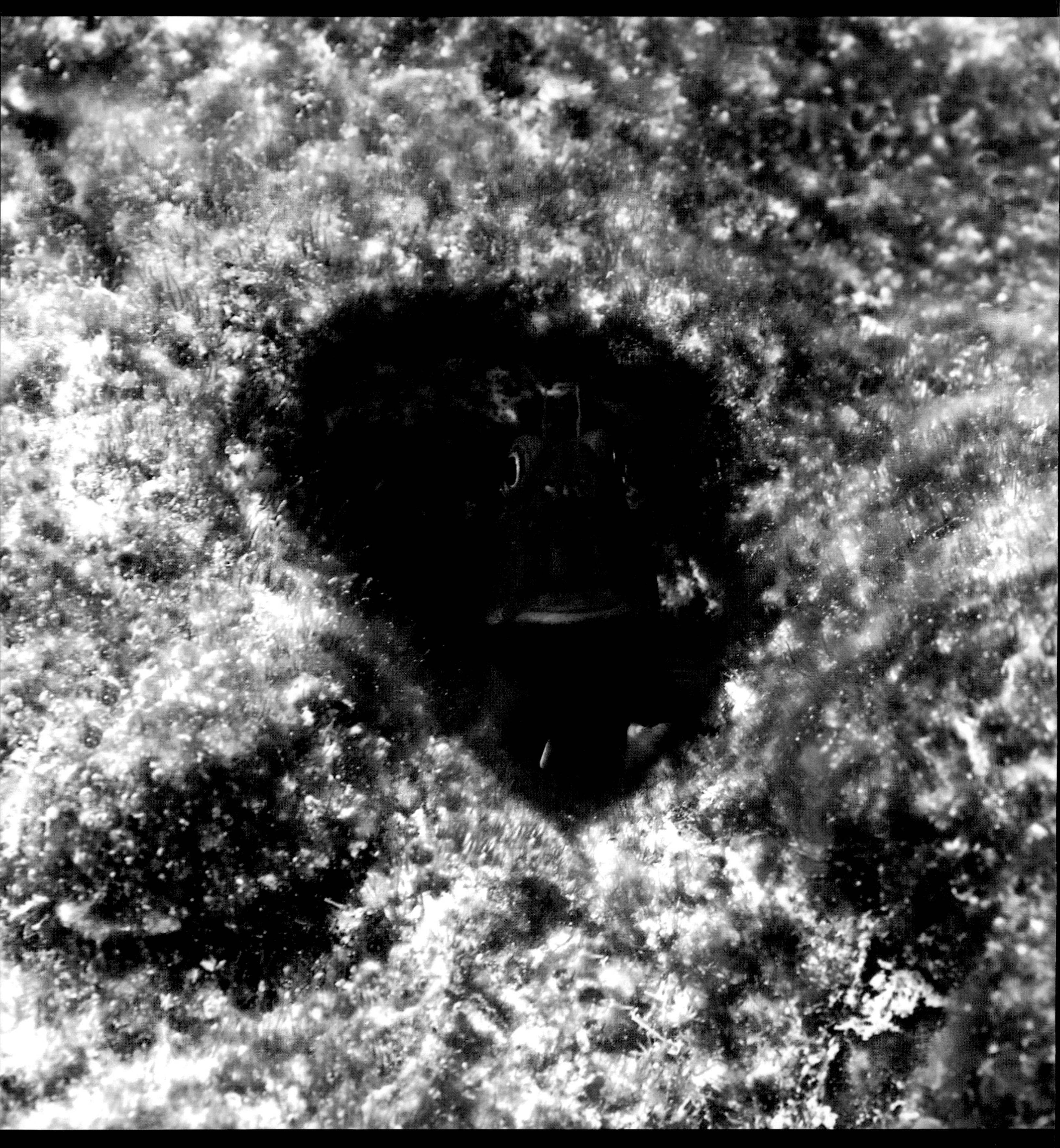

Red lip blenny at home.

*Trumpetfish in the lambent copper light.**
—CF Wm. Faulkner, *The Light in August*

Blues trumpet: I been warm so long it feels like cold to me.
—CF Richard Farina, *I Been Down So Long It Looks Like Up to Me*

My, Snorkel Bob's, poor old gray-haired mother used to wrap things up in a strange rendition of Ma Kettle meets Mae West by saying *Well, asshole she wrote*. Mother has impeccable manners but still dabbles in the vernacular, just for fun.

Snorkel Mom at 90 with great grand barracuda, juvie females

I am reminded, because I think asshole she wrote for this visit to St. Croix. I.e., we're winging west once again, this time to...uh... How about, er… I got it! Someplace tropical! With reefs!

But first, a sweet bye'n bye from a few new friends:

Profile in courage, this hefty cuda is apparently tourist tolerant, so far.

Behind every great yang is a sweet yin—hummingbird at Cane Bay.

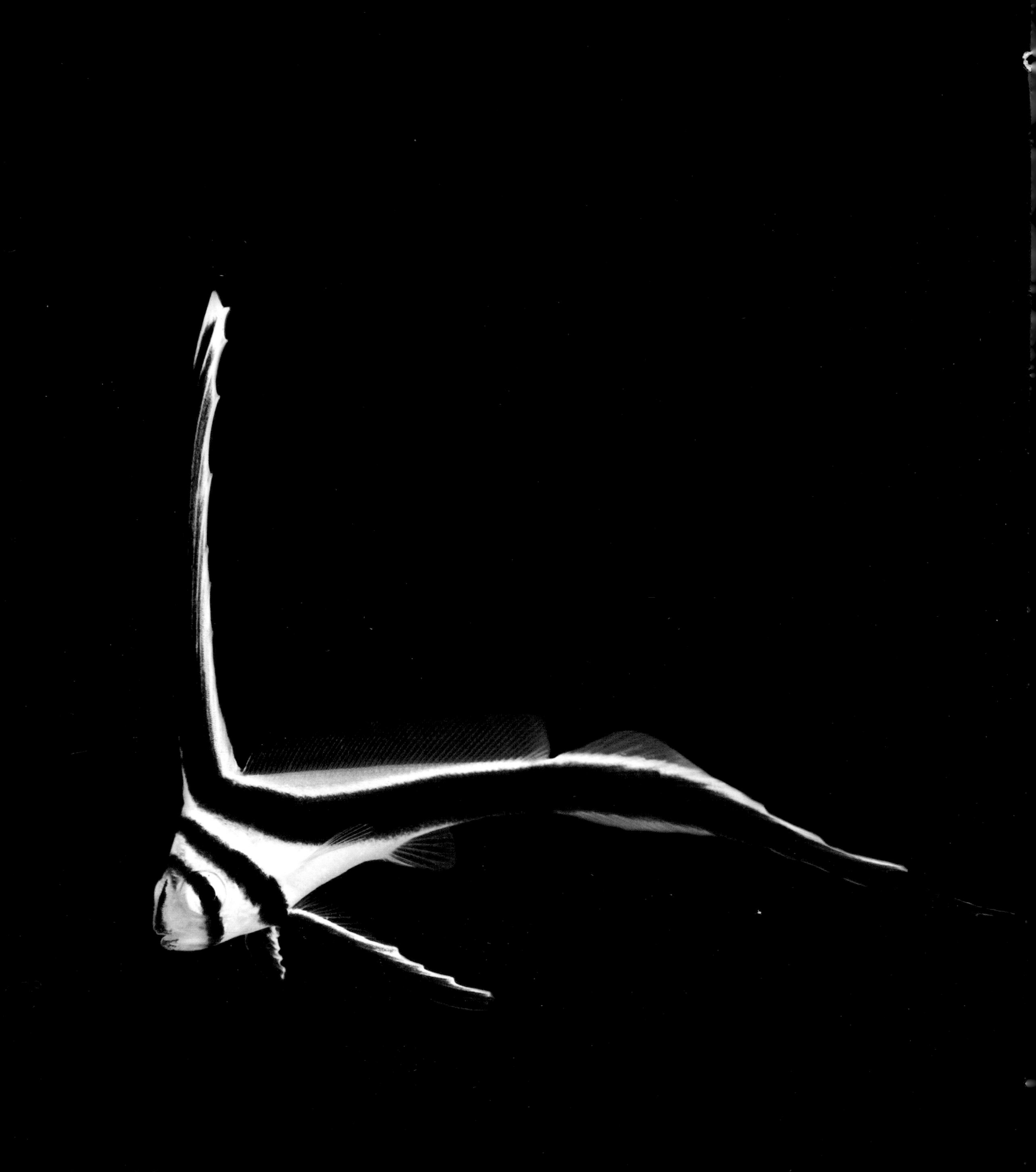

Now we hele to…

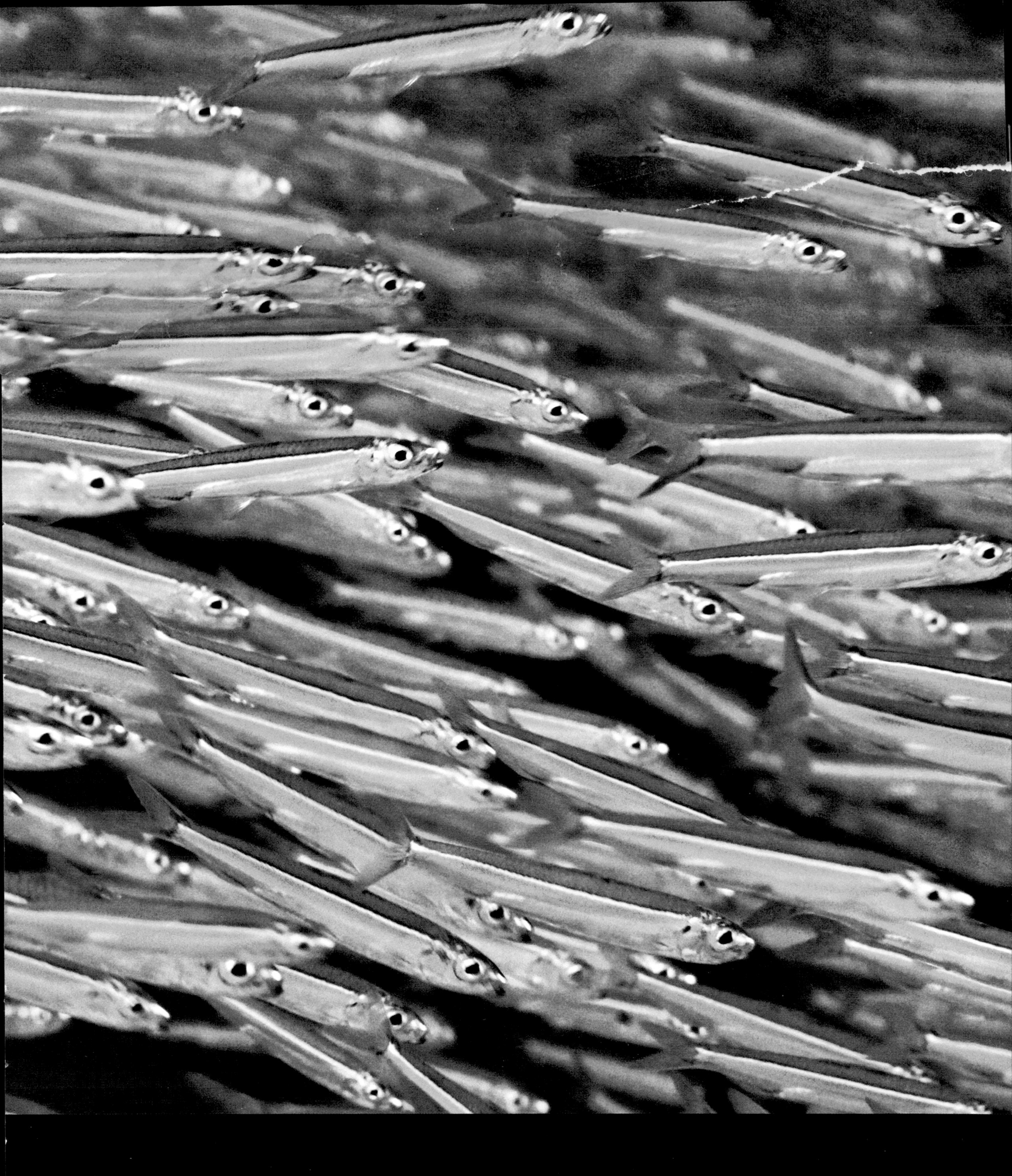

Come on, come on, come on, come on, come on…

TAHITI

Where fringing atoll protects shallow lagoons from breaking waves, providing homes to millions of reef critters ready for the family album in available light. Snorkeling, that is—an immersion in free-form liberation with mask, fins & snorkel. Just add (salt)water and a few dancing sunbeams to dapple the coral heads—oh, and some reef citizens coming forth in curiosity.

Bienvenue, mon ami...

Lemonpeel angelfish tantalize and tease as few Tahitian beauties dare. With coy flirtation and turquoise eyeliner, they pose perfectly from a distance but skedaddle on a distance closed, most often with molten brilliance to blind a focus module every time. What can you do? Try again. As in any romance, sometimes you get lucky.

These angels live in Maharepa, just around the corner from Cook's Bay on Moorea.

Fringing atoll is formed when the volcanic structure finishes its emergence and begins its descent back to the depths from which it came—leaving its peripheral edge just above or very near the surface of the sea. This action takes millions of years.

Meanwhile, the scuba crowd goes outside the atoll to open ocean to attain depth and see bigger animals, like sharks or rays. A well-known atoll in French Polynesia is Rangiroa—pronounced Rangi-rwa by the Tahitians—with only a few square kilometers on the periphery inhabited by humans, Rangiroa has no land in the center, but a lagoon spans 80 miles by 50 miles. Rangiroa totally submerges every hundred years or so, which could refresh and rejuvenate much of human habitation. Think of the jobs created in rebuilding the strip malls and subdivisions.

Fringing atolls have openings with dramatic tidal current and fishy migrations. Rangiroa is famous for its massive shark and ray migrations in very strong current in November.

Adventure potential on the outside is monumental, though snorkeling inside an atoll can be the best in the world. Imagine, if you will, acres of staghorn coral prolifically populated by reef species at 4' to 12' deep. Or

shoreline reefs casually inhabited by lemonpeel angels, pennant bannerfish and dazzling chromis. Oo. La la.

I, Snorkel Bob, hate to shoot my proverbial wad on the mad color issue, but Tahiti reefs recall a feeling from decades past: we just knew there was plenty mo where that came from.

Above is the chromis queen in a rare public appearance, and below is a chromis green—often integrated with blue and turquoise chromis on the same coral head, going virtually invisible when rising into the water column. Each finds her proper niche in the spectrum, I think, unless too much Tahiti weedy goes to purple haze. The tropics can get tricky that way.

Outrageous color mixes freely with unlikely shape and form, which brings us to another dark interlude, the death of Opanohu Beach Park. Oh, the humans of various shapes, sizes, and colors still languor on the grass in the shade and gentle breeze. Some may sense the curative power of a tropical balm far from home while others resent the crowding and wonder if their elite status, *Born & Raised*, is plain to see.

I, Snorkel Bob, was born & raised on the Ohio River and grew up with small ducks and baby catfish, mantis chicks, tadpoles, salamanders, and many other juvies born & raised nearby, who took to adolescence in my care, when I wasn't across the river betting on the horses with my old man or buying firecrackers legal in Kentucky. Tropical natives will never know these things, and now these things are gone, paved over with habitat for humanity. But I cut the tropics boys some slack, just as they tolerate me, some of them, as we slog onward in thickening density.

Opanohu Beach Park was a jewel in the crown of Moorea, sitting at the top between Opanohu and Cook's Bays, a reef resplendent with healthy coral and critter minions. You may remember a most lovable portrait in *Some Fishes I Have Known*, initially entitled *Not a Baboon's Ass*. That was a joke at the expense of a regal octopus willing to sit tall for the camera.

R.I.P.

It's gone, as in dead and gone—gone the octopus, his lair, his coral head, and surrounding reef. It went to rubble, about 98% of it in 18 months. A few solitary critters linger, perhaps wondering what happened and why. Of course perception is selective and contextual, so maybe it was only me, Snorkel Bob, seeing confusion and uncertainty in a wayward fish:

Emperor angel finding cold comfort on a dead reef.

But the threat posed by this reef triggerfish was neither imagined nor tentative. Maybe he was guarding a nest, or maybe he felt rage at the desecration—charging in, shaking fiercely.

He's called humuhumunukunukuapua'a in Hawaii, but we can call him plain pissed off in Tahiti.

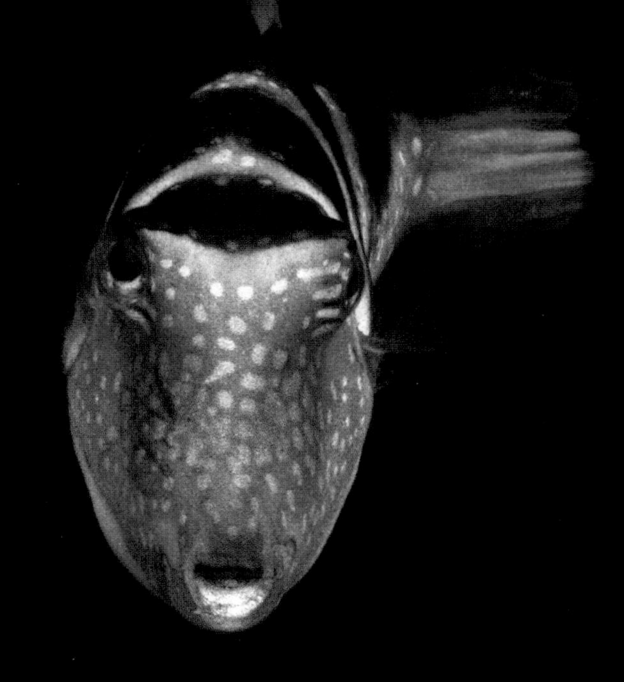

Neptune speaks: *What have you done?*

How many beached humans at Opanohu Beach Park know what passed before their eyes?

Opanohu snowflake eel adapting to reef rubble.

I, Snorkel Bob, advised many years ago that friends don't let friends snorkel drunk. But this may be the moment to raise a glass, to reflect on what passes before our eyes too. Here's to Opanohu Beach Park and the fish people driven from their homes. Here's to saving what we can.

It's down the hatch and around the coral head, because John Lennon advised just as many years ago that life is long, and we can't boo hoo hoo the good times too.

> ...because equilibrium in nature is also social and tempered by understanding. Happiness and balance are evident here under foreboding yet sunny skies. This shallow cove inside the atoll is lively with hungry children, or so the fish seem, roiling the surface in playful anticipation. A thick, gold-plated—that is to say cheap—watch, mostly worn to pot metal by friction and salt air, sparkles sparsely, as does the man.
>
> The darkest clouds converge in a squall that rushes in when he's halfway out the long pier. He tosses bread hunks to excited schools who boil the surface. Oblivious to the downpour or encouraged by it, they recognize the familiar face, and they frenzy in simple pleasure. Without looking up, the caretaker pulls and tosses, hobbling to the end, where two more baguettes under his arm let him continue. Grinning intermittently he hurries to finish before the bread is too soggy. His pets churn the water, telling him where to throw next, proving that fish can learn and teach, can know and see beyond the surface.
>
> Just as quickly the squall passes. The gill breathers' spectrum of yellows and reds, blues and silvers, stripes, bars, arcs, spots and squiggles of dazzling incandescence and aquamarine flash in the morning sun. The hotel guests flash an equally garish and gooier spectrum at a more civilized pace, comfortably accessorized with their Danish and coffee.

—from *Flame Angels*, a novel of Oceania

At a very similar but far more frugal establishment in Cook's Bay is a brand new pier running out to where the reef drops off. Construction methods preclude mention of the hotel by name; they stomped the reef to build the forms and pour the concrete, as if the reef was expendable, or maybe it would recover. It's not. It hasn't.

Yet just off the end, perched on a discarded pier block sat Scorpio.

Nohu—scorpionfish

He's likely a scorpion fish but could be a stonefish—one makes you severely ill and the other kills you, which, at the moment of impact, may seem preferable. Both are *Nohu* in Tahitian, and the only way to tell the diff is to count the (venomous) dorsal spines. One has eleven and the other thirteen. But check out the *shayna punim* on this pup—you have to step on them to trigger the venom, which you'll likely trigger in anyone you step on.

Nohu puss up close.

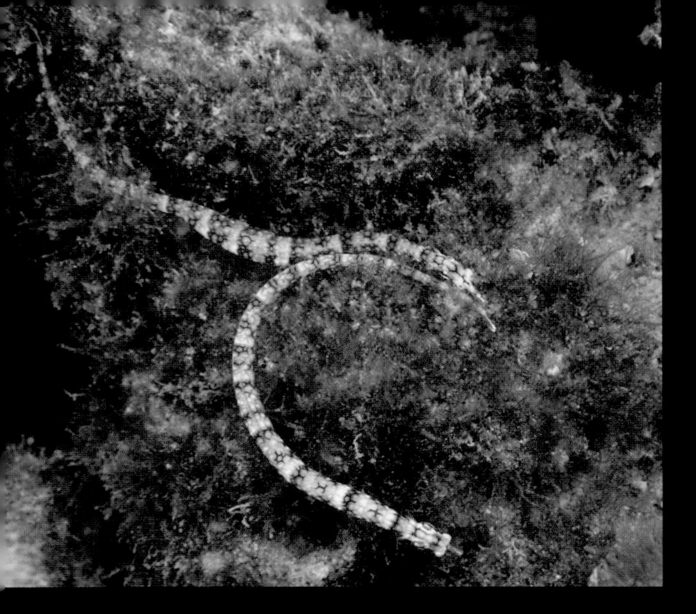

On the way out to visit Scorpio are coral and vegetation in very shallow water, including a gold-brown, fuzzy plant attractive to pipefish. These amazing, delicate critters are 1st cousin to sea horses, here in 10-15" of water—not even enough room to sink the camera.

Pipefish, filly and brood stallion

Go snake-eyed and you'll see the swelling in the pipefish on top—a pregnant male. Female pipefish, like female seahorses, deposit eggs into the male's brood pouch for fertilization AND gestation till birth.

This shoal reef on either side of the dock actually emerges a few inches at low tide. High tide offers marginal flotation—and that's only for the wispiest of mermaids, who can meander the rocks and coral to reach the sandy dips where a critter might take 5 to catch some rays in repose.

Synchronously as things often unfold, a meeting on a shallow reef between an octopus and a mermaid presented itself for development on these pages some months later, on the same morning that a story appeared in the news. That is, PETA (People for the Ethical Treatment of Animals) is up in arms, so to speak, against restaurants specializing in live fish, particularly octopus that can be eaten as it squirms. Nature most often delivers death quickly, and no other predator on earth has so overwhelmed nature for its own convenience.

Eating creatures is a personal decision. Eating creatures taken from the wild is a different decision that should take into consideration severely shrinking natural habitat and burgeoning human population.

While a human person with a speargun might think of lunch or peer-group praise—or a Manhattan pedestrian might fancy a *de rigueur* amusement in a chic bistro—Octoanita pondered formatting and focus for a session of memorable exchange.

photo by Anita

And don't forget the love, effusing here in Valentine shades. This octo posed in barely 3' of water. Flashing went red to orange to red-orange to brown to green and back again to red—perhaps signaling hope that the octopus shall endure for a while longer.

Just past Maharepa on the way to the airport is a Sofitel Hotel with a sheltered cove that is a Marine Protected Area—protection from extractors does not protect a reef from runoff, effluent, and carelessness. But so far, this place is A+. It looks like sandy bottom from a distance and from the shoreline too, with isolated rocks, and out ¼ mile is some beige stuff, maybe seaweed.

Except that the isolated rocks are small coral heads teeming with critters, and the beige stuff is antler coral—acres of it thick with many individuals of many species, some of whom personify the convergence of strange shape and brilliant coloration. Personify? Nudibranchify?

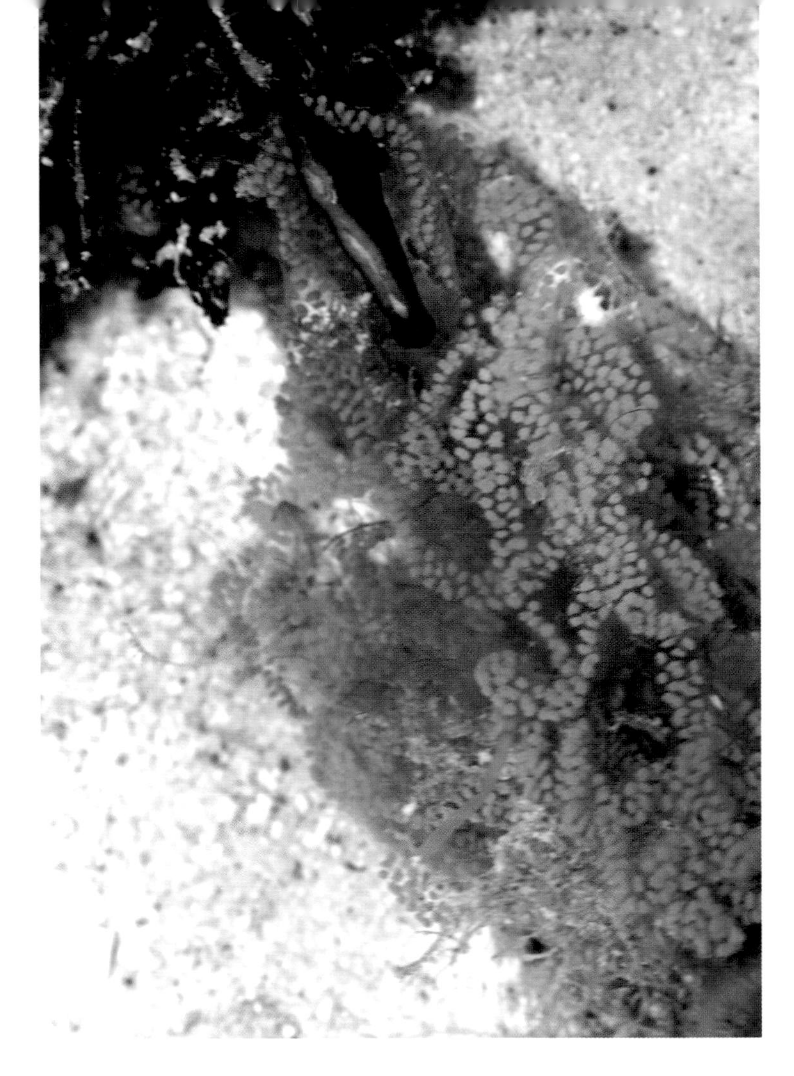

Nudibranch egg string

This stuff looks like eggs. Nudibranchs are always flamboyant, usually uninhibited, and generally poor at family planning.

photo by Anita

What looked like sandy bottom from the road was actually grassy—with critters feasting on a windfall crop of nudibranch eggs. Beyond the sea grass comes ¼ mile or so of sandy, rubble bottom with intriguing traffic that included…

Moorish Idol, a species unto itself.

He's called a guineafowl puffer in Tahiti, same as a spotted puffer in Hawaii. He's in the sand flats at Sofitel on Moorea, but not too far from cover.

The *bluestreak gobies* (below) are often seen in pairs, often guarding a burrow in the sand. They showed up in St. Croix too, though their only appearance in Hawaii for me, Snorkel Bob, was in a pet store, ON SALE for $15! The Maui County mayor had a "buy local" campaign, and wouldn't you know it; the pet shop owner showed up for the photo op. On meeting Ms. Sue from Snorkel Bob's in Kihei, he said, "Oh, you're the guys trying to put me outa bidness!"

Said Sweet Sue, "Nay, Sir, it's you putting US outa bidness!"

Let the aloha begin on the memory of our late, great President of the United States of America Ronald Wilson Reagan, who best summed up our hopes and dreams when he said, *Mr. Gorbachev, tear* down *this aquarium!*

Must a pet shop sell reef fish? No, it mustn't.

Threadfin butterfly

Why would Neptune make two fish so similar yet different?

Lined butterfly

Also cruising the sandflats, a *Scrawled filefish*

This big boy went just shy of 3'—NOTE the effect of sunlight in the shallows, say 5-8' deep. Shooting down against the white sand, a careless photog can damn near fry his shot, while mere moments away, shooting above midrange into the water column, this *Moorish Idol* and *triggerfish* are cool in the azure blue.

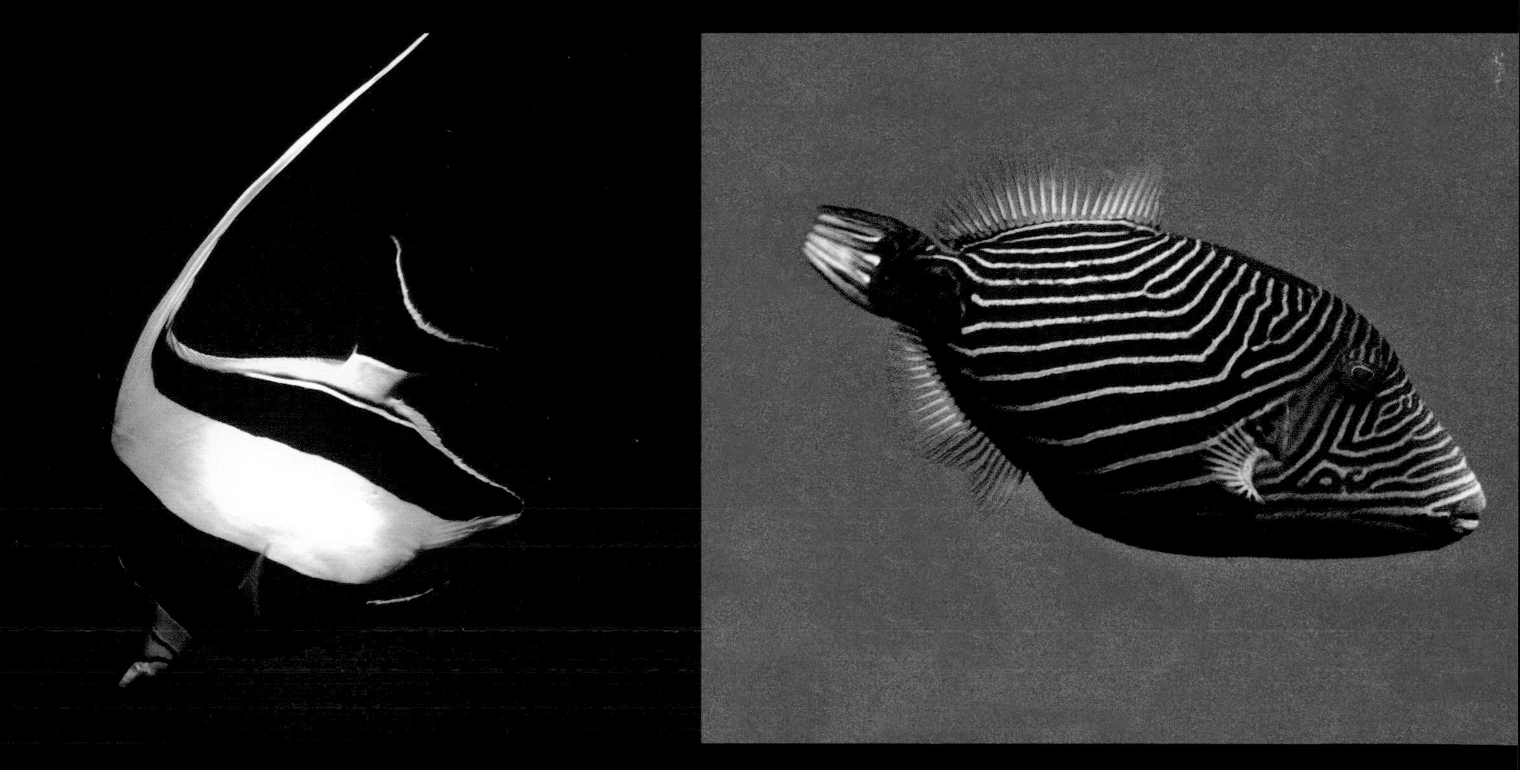

Wait! Trigger is telling us something…

He means here at the staghorn coral reef, the approach littered with dead coral and rubble, indicating harsh current or breaking waves.

Saddleback butterflyfish, once plentiful in Hawaii, are now sold on the Internet. Casual encounters in person occur mostly in Tahiti. The staghorn coral reef at Sofitel on Moorea begins in earnest, teeming with damselfish…

…with cardinalfish…

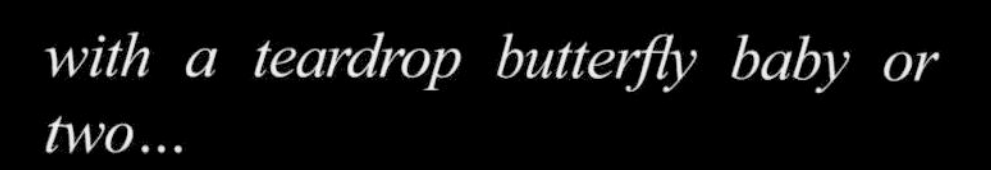

with a teardrop butterfly baby or two...

...parrotfish of many colors...

…an epaulet soldierfish…

…and, of course, much much more.

Then it was time for a visit to Tahiti—the island of—known for urban density, traffic, grit, litter, and other assorted challenges of the human drama. Yet with open hearts and minds and some website info, we visted the reef at Taaroa. It is failing, but we must enjoy, lest we spend our time on Planet E missing the last vestiges of the Spirit of Our Time.

The lagoon inside the bounding atoll at Taaroa goes for miles with intermittent distress. A snorkeler may sense that the end is nigh but can still have a memorable outing here, encountering what can only be fantasized in northern climates—soon to be fantasy in any climate. Beyond coral mortality of 20-50% on any given section of the lagoon, the dark influences (effluences?) bearing down include night spear-gunners. These reef killers destroy at night with scuba tanks and flashlights by wiping out the parrotfish *as the fish sleep*. Remember the Carolina parakeet? The American buffalo? It's happening again, underwater and out of mind.

Any marine biologist not employed by a State agency or a corporation will confirm that parrotfish are a lynchpin species. When the parrotfish go, the reef collapses.

Reef collapse is occurring in Taaroa. Observing the carnage—the flashlight guy out yonder at 6-10' depth at night—I, Snorkel Bob, asked how the people of Taaroa could decline to protect their reef. How can they allow this? A woman born (& raised) there has watched it decline steadily. She shook her head, hardly sanguine, deferring to the folksy platitude for keeping the peace, that "there are plenty of fish for everybody. Everybody must be allowed to fish."

Therein is the problem. The fish can't keep up. Humans are taking everything.

The crown jewels are not yet entirely lost, and the task of salvation may fall to the next generation of human and fish children.

Are you my daddy?

Though healthy reef is sporadic at Taaroa, some isolated bright spots like this coral cluster host several species, indicating balance.

Coral feeds and filters while providing habitat and temporary cover, in case a fish wants to hang out and reflect

on what might be done.

This Taaroa butterfly looks surprised...

...while just overhead a citron butterfly grazes with nonchalance.

As with all reefs bordering human habitat, farther out means less dead stuff. Out near the atoll barrier, a small dascyllus comes forth with confidence.

Big swatches of mossy rock rubble characterize the lagoon at Taaroa, and though these areas appear to be compromised—brown—they host fishes of remarkable colors.

6-bar wrasse

Portrait of the Artist as a Young Wrasse

Striped surgeon

An abundance of gregories is another symptom of life as we like to see it. Hardly as flamboyant as his neighbor, he's nonetheless resplendent, framed in delicate mosses.

Then it was time to go. Coming into shoal depth of 10" or so, we met the most playful pups I, Snorkel Bob, have seen in a long time. These two dragon wrasses or rockmover wrasses are about 2-3" long and possibly litter mates, except that they don't come in litters, not in leaf piles under the house or baskets with fuzzy blankets. They come from fish eggs. So why the friendship? I don't know, but a couple of tumbling, frolicking rascals never had so much fun in the shallows of Taaroa Lagoon.

Okay, where next? Well, it's a loaded question and you musta seen it coming. It's off to the land of Aloha to check in on the home crew.

It's always good to get back to the old familiars, but it's tough leaving new friends behind.

Wait! Snorkel Bob! Don't go-oooooooooooo….

What is it, Tobyboy? We got a flight to catch. My back hurts and my hernia is acting up from schlepping all this gear and frankly, this itchy sumbitch on my…well, never mind. What…

Oh! DAMN! I nearly forgot! Okay, we'll duck back in to say hey and get one last shot to keep the home fires burning in our Tahiti hearts. It's all one reef, boys and girls. If you don't love it, you may need counseling…

Oh, don't look so glum—you'll likely need counseling anyway.

Hey, who'da thunk a pennant bannerfish would come in close for a portrait?

Goodbye pennant bannerfish! Goodbye Moorea.

Now let's hele on, boys and girls. It's…

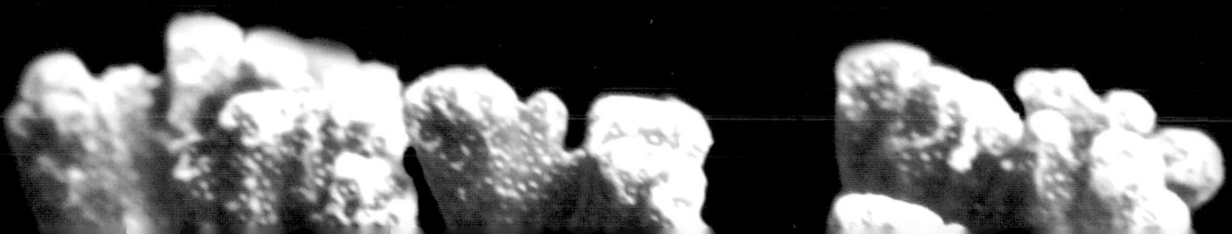

HAWAII TIME

Where you can get so pissed off at the government you got steam coming out your ears, but it's okay, because you can go to a public hearing and get a load off your chest. You get three minutes to deliver the facts—AS IF they're not crystal clear.

Your testimony will feel like a tree falling in an earless forest, but you'll see some friends and maybe get a few brewskies after and everyone will feel better because misery loves company and maybe you'll plan a snorkel outing for tomorrow and you'll still be pissed off but not as bad.

Boy oh boy.

Okay, where we oughta go?...

T-H-E most photographed fish on any Hawaii reef is the hawkfish, so named for his hawk-like behavior of perching in stillness, waiting to swoop on smaller prey passing below unaware—an ambush predator. He shows up in many photos because of the stillness. Sumbitch doesn't move, so he stays in focus often as not. If you only have a few days on the reef, and your shots are looking iffy, you need to find some hawkfish.

This model is the arc-eye hawkfish for obvious reasons. So I, Snorkel Bob'm cruising past this shrimp of a hawkfish as one teency tiny thought bubble fills with *Boy, anybody gets another hawkfish shot oughta pay a surcharge.* In the next instant Igor here opens wide in a ferocious yawn with his pecs flexed and his dorsal in maximum flair—

What the!

Proving that presumption is not humble and may verge on arrogance. Is this half-pint a giant among fish or what? Which further underscores my contention that traveling far from home is great for exotics, but it's not the exotic species that count most in fish or any society; it's the individuals and the behaviors any one may display at any time.

Hey, the shot to the left was a cakewalk, a snap in passing. But don't most things done well look easy? Just for fun, I got Igor again on a different angle to show his consistency—to show what a character a tiny fish can be.

And before you know it, presumption turned to humility. Spectacular hawkfish demand a presence in any record of Neptune's glory.

Blackside hawkfish

Any fish can have a unique personality for that matter, and here's another milestone on this tour of reefs and some things a fish might convey. I, Snorkel Bob, have read opinions on the "sentient" capacity of fish. One chat-room fellow claimed that fish may be sentient, but they're not "self-aware." The next fellow said sentient means self-aware. I, Snorkel Bob, think the first fellow lives in Michigan—no disrespect—and the second fellow lives on a university campus.

Here's my contention and the point of this tour: given repeated exposure to humans—humans without nets, spearguns, hooks, clubs, or other weapons of fish destruction—fish will warm up, socially speaking. All this bruhaha about self-awareness flies out the window when you get to know a fish, when the fish sees and recognizes a familiar being who may be sentient, self-aware and compassionate, who may come with only one "fishery" in mind, which is that of contact between species in which neither must die. Let's break down this Hawaii section geographically and begin with a reef where no extraction is permitted, a Marine Protected Area (MPA), to see who has what to say.

olokini Crater is an islet, a remnant rim of an ancient volcano between Maui and Kahoolawe in Malaaea Bay, west of Kaiwe Channel. More than ½ the rim is gone, the other ½ rising 100' or so from the sea with habitat for some hardy grasses and a few shore birds, scissortail frigates among them. What's left of the crater rim dips into the sea at each end, each a popular reef with unique characteristics and fishes. Between the ends, the crater bowl comes up to any depth a snorkeler might favor.

Back in the day, Molokini had no moorings; boats anchored on the bottom, often in coral. A proud moment for the Molokini fleet was in 1984, when the Navy discovered unexploded bombs in the crater and gave notice to STAY OUT, because the Navy planned to detonate for the betterment of the citizenry, albeit the ruination of the reefery.

The navy bombed Kahoolawe for decades as target practice, till Hawaii cried foul—not the State of but a few Hawaiian people, practitioners who woke up from cultural coma to stand up and speak up. Meanwhile, some Navy pilots had bombed Molokini instead, because it's only a few miles away and may have appeared in need of a bombing. Oh, those boys and their bombers.

The bombs in Molokini Crater may not have been live but would be blasted to Kingdom Come nonetheless—until a few intrepid fellows dove on the bombs the night before. The night divers fixed tow lines and towed the bombs out to very deep water and released them. The Navy was in a dither on account of the destruction those bombs could cause.

The reef at the southern tip is Enenue, a rich and variant community of fishes and invertebrates, like this fairy or blue dragon nudibranch. It's not really a reef centipede, but it looks and moves like one, and it eats hydroids, so yes, its cerata will also sting like fireworm.

The reef at the northern tip of Molokini Crater is Reef's End. The rim extends for ¼ mile or so underwater at Reef's End, sloping on both sides and shallow on top, say 6-12'. Outside can churn a current neither you nor the Olympic wunderkind can beat, but on calm days the outside wall is busy as Grand Central Station and probably safer. The end of Reef's End drops off quick and deep and is a good place to occasionally see big animals—whale sharks, other sharks, manta rays, and whales—at any depth.

These two critters are snorkel shots:

photo by Anita

photo by Anita

Inside the ridge is no current, with a slope to a sand channel notable for the garden eels who live there—and also notable for some of the fishiest neighborhoods in Hawaii. These fish know tourism. Let me introduce you to a friend of mine: Razorboy, resident host of the eel garden flats in the sand channel of Molokini Crater. Extroversive to a T, this maitre d' runs the show. He's a peacock razorfish, in the wrasse family but hardly irascible.
Table for one? Right this way.

Razorboy approaches the camera like a natural—or is he approaching me, recognizing an old friend? I think it's a little of each, as he makes a great show of sifting some bottom schmutz very nearby or turning to advantageous profile while keeping a pop-eye on me.

Yo, Pauly told me: it's Malabar in the 8th, on the nose.

Razorboy makes his rounds.

From the family album archives, it's Razorboy as a tiny pup! (I think.)

Also thriving in the sand channel, many red-spotted sandperch, aggressive feeders whose population indicates a healthy and balanced system.

I might sound crabby, but this is the 2nd hermit I've seen on Hawaii reefs.

The aquarium scourge relieved Oahu reefs of 300,000 hermit crabs in a single season and sold them to mainland resellers at 11¢ (ELEVEN CENTS) each. Now they're gone—a species vanished from that reef system for a gross return of $33,000. What would it take to restore a species? The government studies alone would cost millions and take years. Hermit crabs are so much fun to watch in an aquarium, so busy, so nosy, so obtrusive in their pursuits.

Here's the damage: as they grow, hermits change shells, leaving the old shell behind for the bigger shell. When a hermit is taken, so is the shell. Many reef experts consider hermit crabs a lynchpin species; without the hermits the reef will collapse. The aquarium gang now takes hermits from the Kona Coast, a bigger cache to wipe out. The aquarium hunters also cry foul—*but we don't take so many hermits anymore!* Duh. As if the aquarium hunters are special and can catch their hermits and have them too. They've enjoyed non-detection for a long, long time.

He's a juvenile barred filefish, unseen in your reef fish ID books, so this shot is rare. These fish are known to suffer juvenile inferiority complex because of so many skin splotches—you remember the drill, when squeezing only made it worse, and Clearasil made it look like a patch job on some damaged sheet rock. Not to worry, they grow out of it, gaining social confidence and poise to compete successfully in the reef dating scene.

Hi. I'm Steve.

I, Snorkel Bob, did not make that up.

Wait a minute. Who's this?

It's a cornetfish. I call this guy Slim. I call all cornets Slim, and you can see why, as Slim turns for close-contact cruise-by.

You can call it *manini* (a trifling), but I sense the cornets coming closer at Molokini. You can quibble that proximity is not engagement, but I, Snorkel Bob, think it's a step in the right direction. Proximity indicates greater trust—less fear comes from repeated encounter with no death, no capture—and therefore a greater sense of curiosity and sharing, the roots of society.

A fellow I know calls himself a fisherman and calls certain species "stupid" for allowing him to approach and spear them. He's a gross, overbearing individual whose taste is in his mouth. He loves glazed donuts and frothy cakes and gained 40 lbs in the 3 years I've known him. He further distinguishes several species of sharks according to their position in his "good eating" assessment, and I can't help but wonder how a "sentient" guy is so blind to what's before him. I certainly hope he doesn't get ciguatera—an organism in coralline algae that accumulates from the bottom of the food chain to the apex predators, mostly big jacks (ulua) and sharks. Harmless to reef fish, it makes humans itch intensely for two years and then die. Have a nice lunch.

Wha? Wha'd I do?

While we're in the sand channel at Molokini Crater, we'll cruise over to the coral ridge thicker with fish than most. The coral on the ridge is dead, so the only attraction I can speculate is social, including a few busy Hawaiian cleaner wrasses working profusely to meet the demands of a vibrant reef population—protected from extraction.

Get the oogies off my scalpel, Cleanerboy!

I got a goober in my gill plate, Cleanerboy.

Shame came to the State of Hawaii in the Linda Lingle administration (R) with a Chief Policy Advisor who was a wholesaler to the aquarium trade. As we speak, the State of Hawaii allows unlimited export of the Hawaiian cleaner wrasse, a fish found nowhere else in the world, guaranteed to die in days in captivity, when it could be working to protect Hawaii reef fish from parasite infestation. The lead (political) scientist at the Kona Division of Aquatic Resources office calls aquarium collecting "the most lucrative inshore fishery." That's an economic assessment, a data spin to prove a new hypothesis. This bureaucrat's hat is inside out, or maybe he too needs a glass navel. Can you tell the difference between science and "science"?

Here's the real kick: the Hawaii State employee referenced above is the most ardent, outspoken advocate of aquarium extraction in Hawaii. It's his baby, and he assures hobbyists across the (main)land that they shouldn't worry about yellow tangs going to $100 each, because he has things managed. Yellow tangs now retail for $40-50 each, though each leaves $3-4 in Hawaii.

I, Snorkel Bob, think he has many people fooled, and we're working to change that. For starters, the most lucrative inshore "fishery" reports annual revenue statewide of $2 million. **Reef-based tourism generates $800 million annually—400 times greater—and they're killing it!**

Adding criminal behavior to moral failure, aquarium collecting is managed as a "fishery" in Hawaii. A "fishery" seeks optimal revenue on its catch. But the laws are far different for wildlife captured as pets than those captured as food. Humane treatment makes predictable mortality a crime; if we know the animals will die, their capture is illegal. The Hawaii cleaner wrasse must have 30-40 other fish to clean, or it will die. When I, Snorkel Bob, visited the biggest exporter of Hawaii aquarium fish, I saw tanks with Hawaii cleaner wrasses.

"They'll die," murmured I, Snorkel Bob.

"They're fun," replied the Hawaii reef fish exporter.

A goldring surgeon, or *kole*, gets cleaned by a Hawaii cleaner wrasse above, so he may stay healthy to perform the same service to a *honu*—Hawaii green sea turtle (below).

THAT is reef balance, reef health, reef sense, with no crimes committed. Pardon me, Snorkel Bob, for getting so lathered up at the mere thought of reef plunder, but time is running out. I discussed the situation just this morning with a woman in Washington, D.C. who said, "Oh, my, it's so terrible. You know, I was at my accountant's office and he had this aquarium. He only had 4 fish in it, big ones, and I was going to say something, but I didn't. But I thought, boy, it sure would be boring to live in that thing." Bingo, lady. Boring and then some.

PLEASE GO VOCAL ON AQUARIA ANYWHERE like our 41st President of the United States of America, Ronald Wilson Reagan, who stood up, stepped up and spoke up…

Mr. Gorbachev, tear *down* this aquarium!

If you see an aquarium in an office or shop or bar or restaurant, tell the management that you cannot spend till the aquarium is gone. If the aquarium stays, then spend elsewhere. Please.

Where was I? Where am I, Snorkel Bob?

Snorkel Bob! Snap out of it! (Thwack!) Get a grip, man! It's okay! You're here with friends in the sand channel at Molokini Crater.

Oh. Hey. Thanks, Doctor.

I call this fish Doctor in deference to his professional status. He's an orangespine surgeonfish, and yes, his caudal peduncle is armed and dangerous with 4 razor-sharp scalpels—big'uns, 2 on each side—in case of a hiney approach he may consider inappropriate.

Okay, let's see here. We're at the north edge of the sand channel near Reef's End at Molokini Crater, where your wrasses congregate. I, Snorkel Bob'm painfully aware of my juvenile inclination and arrested development, but some things still make me laugh, like your wrasses. Why wouldn't they, given their flamboyance and irrepressible good cheer?

Take this bird wrasse:

Bird wrasses keep their distance—and even a sociable bird is difficult to photograph. Something about the deep blue body with red flecks, and the brilliant green splotch above the pectoral fin defies focus. The bird wrasse here was the friendliest I'd met in a long time.

Another unique characteristic of the sand channel's north edge is the preponderance of red wrasses. I'm thinking 1st of the disappearing wrasse, exquisite for its radiant reds and gossamer fins. The many details of color and form in this fish make it one of the most graceful on any reef.

But a reef addict must use superlatives cautiously in profiling grace as a function of color and form, since this same neighborhood is home to many smalltail wrasses:

The male smalltail wrasse (left) is commonly called a pencil wrasse, and his dazzle is apparent, though for sheer, heart-throbbing love the smalltail femme gives me pause every time:

Yeah, I see him.

Oh, Ruby,
look who's here!

This same stretch of the sand channel at Molokini Crater—protected from any extraction—is also home to two species of lined wrasses.

5-line wrasse

8-line wrasse

Hey, what's all the commotion? We're trying to have some grace and beauty here!

Oh, it's the dragons…

This irrepressible character is called a dragon wrasse as a pup, because of the appendages and electric squigglies.

The adolescent wrasse above may still qualify as a dragon, and won't hesitate to show off his rowdy good nature. As he grows, he transforms from a plankton plucker who depends on camouflage into a bold, aggressive hunter who doesn't shy from a camera…

And he isn't afraid to move a rock or two—hence the new name: rockmover wrasse.

Natchurlly, this action attracts followers, and a fish who hardly ever misses a food fray is another gadabout…

The saddle wrasse isn't everywhere all the time; she's only everywhere she might snag some stray snax—bringing us to another PAUSE in programming for a moment of recognition and kinship. I.e., I, Snorkel Bob'm often profiled as assertive, obsequious, and obtrusive if not abrasive. Tsk tsk—I learned *ev*erything from my close friends, the saddle wrasses. We're comfortable in our garish color, our flamboyance, and motivation. In your face, up your nose holes, and under your lunch is where we look. Are you a sea turtle or a parrotfish? No problema.

Saddle Wrasse also has a great sense of humor and seems well tolerated, like here, hamming for the camera with her pal Polly, the parrot.

Polly is a female redlip parrotfish and like most parrotfish (*uhu*), she's very easy company. As noted, parrotfish are integral to reef survival—remove the parrots, and the reef will collapse. Parrotfish eradication is legal in Hawaii—not in Hawaiian culture but in the State of Hawaii, where speargunners with scuba tanks or hooka hoses from surface-supplied air can dive at night and pluck every parrot from a reef. The parrots will be eaten in a few days and will not return to the reef for years. This criminal behavior has been a topic of "scoping meetings" around Hawaii for years, orchestrated by the Department of Land & Natural Resources.

Redlip parrotfish—though critical to reef survival, are methodically eradicated from many reefs in Hawaii with no protection.

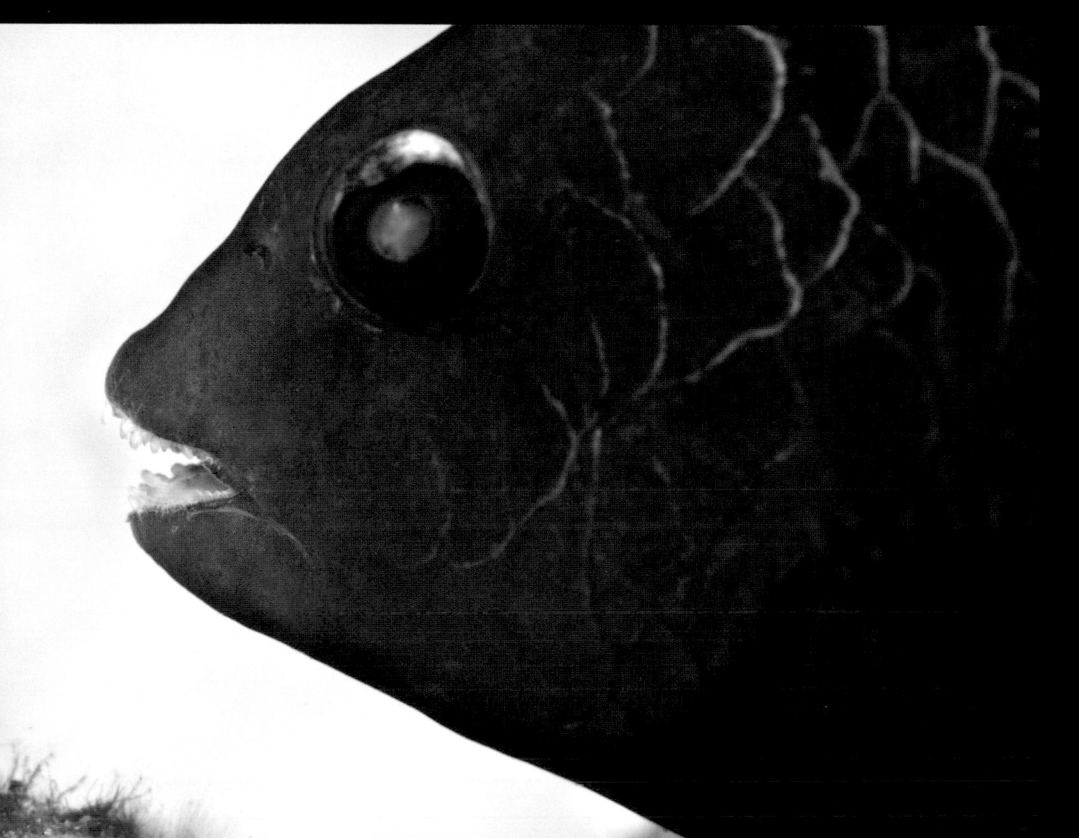

Another parrotfish you may only see in a Marine Protected Area is the stareye parrot.

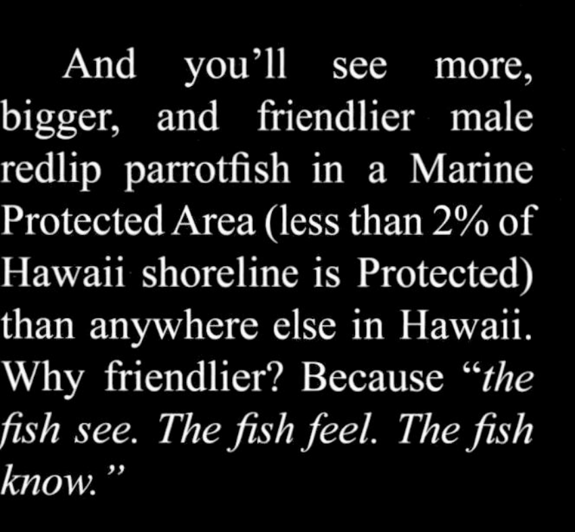

And you'll see more, bigger, and friendlier male redlip parrotfish in a Marine Protected Area (less than 2% of Hawaii shoreline is Protected) than anywhere else in Hawaii. Why friendlier? Because "*the fish see. The fish feel. The fish know.*"

—Kimokeo Kapahulehua informing me, SB, on the fishy P.O.V.

Inveterate residents of the Molokini sand channel include many wrasse species unseen elsewhere, because Molokini is protected from the aquarium scourge. Otherwise, the aquarium demand around the world would wipe this slate clean too, voiding this vibrant reef of a 2-spot wrasse in full flair.

He allows one shot before vanishing.

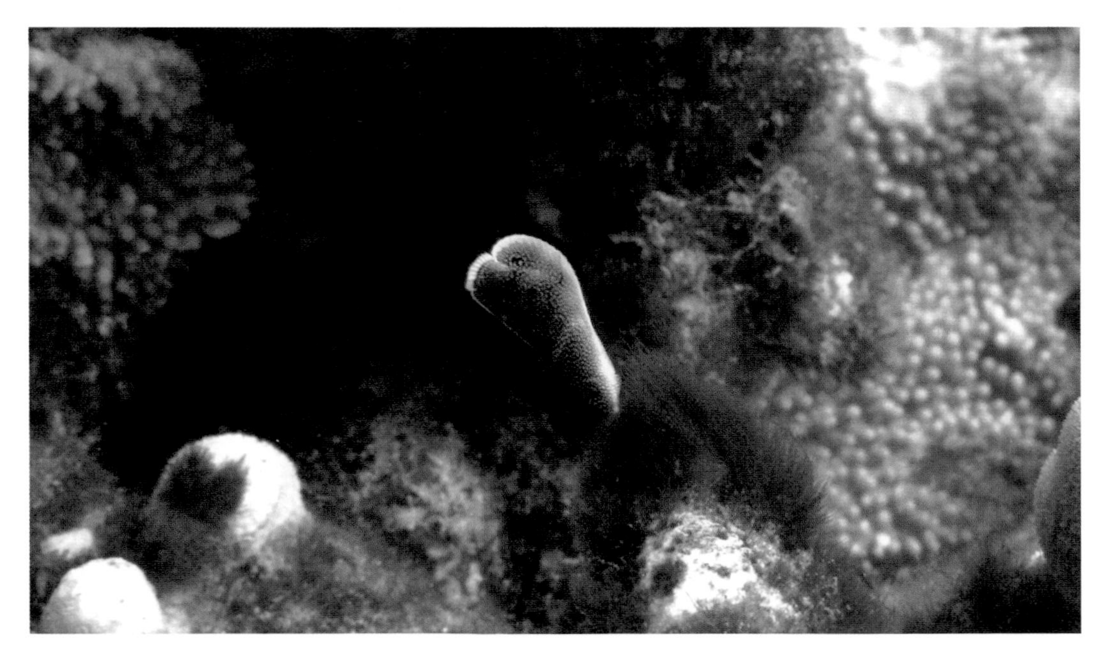

This tiny psychedelic wrasse has far better odds on survival, because Molokini Crater is 100% protected from extraction. The crater also has nearly 100% coral cover—do you think there's a connection? Futz, duh; my reef assessment is based on empirical observation and seems more sound than Dr. Wrongway Peachfuzz's statements of economic "fact."

The belted wrasse at Molokini Crater is protected from the aquarium trade. Most other reefs are not. Another characteristic of reefs left to nature's plan is the presence of generational individuals—fish in all phases of development. The aquarium trade blows smoke up the collective ass on "sustainability" that allows extraction of only those individual$ prime for the trade. We as a society may feel symptoms of emphysema on the mildest flatulation.

For example, these two tiny pups are shortnose wrasse fry.

And these shortnose wrasses are a juvenile (above)—note the coloration—and an adult (below):

All three live in close proximity on a protected reef, as do these ornate wrasses:

Tiny ornate wrasse

Sub-adult ornate wrasse

Ornate wrasse, adult

It's way more wrasses than you'd find on a reef with no safeguards against aquarium plundering—no limits on any catch or the number of catchers and no constraints on endemic or rare species. This reef is abundant and balanced as all Hawaiian reefs were, not so long ago. It's still no swim in the limu, with hall monitors ready to issue permanent detention.

Out front:

Ringtail wrasse—the underbite and prodigious teefs, I mean teeth, match his gourmandize.

While overhead waits further balance, the natural way: *Barak Cuda, Leader of the Reef World.*

Some rightwing snorkelers speculate that this influential fish may have swum in from Kenya. Swam in? Who cares? He's the leader of the reef world, keeping order along with…

Bigeye emperors:

Whudju say?

You talkin'a me?

The bigeye guys mostly hang in the water column still and dazed as stoners, though their bulk and formidable teefs, I mean teeth, let you know they can shred in a heartbeat.

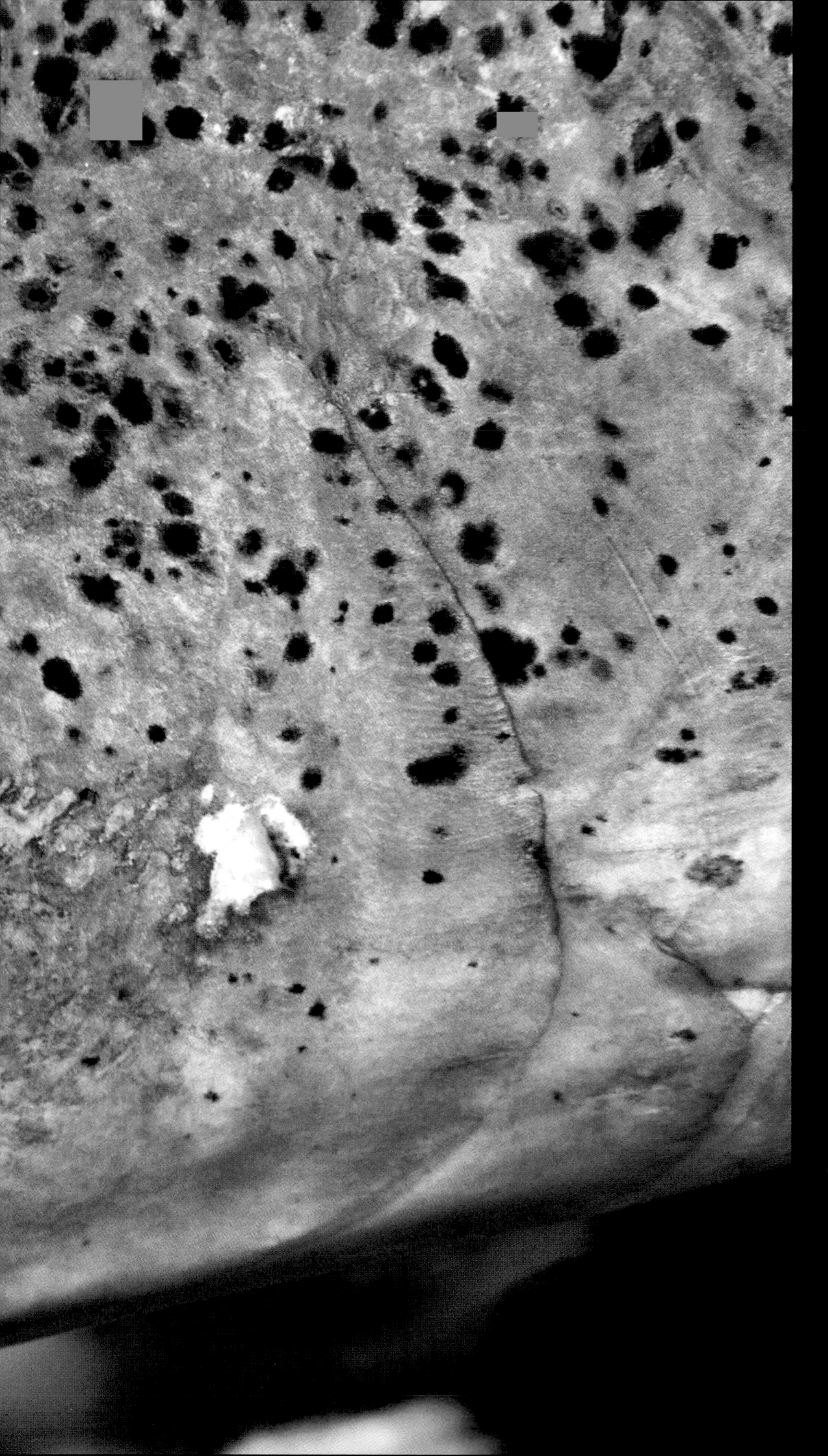

Wanna grab some lunch?

Big Betty here came within kissing distance to offer the invite. Note the facial scarring. It's from ramming her mug into crevasses in a passionate lunge for kissing distance on her prey. But all in all, she's my kinda gal: big, bold, and colorful, a fish who'd rather jump on it than pussyfoot around.

And a most common team working the backdoor flush are *ulua* and moray:

Cruising for burgers—or maybe a fish sandwich.

Hey, Brudda, this whitemouth moray murmured to me, Snorkel Bob, up from his coral crevasse. At 3' and growing, he needs to man up—I mean eel out. Note the adorable row of teefs along the roof of the mouth hole, the better to grasp and tear, my dear, though these pups are commonly curious and sometimes affectionate.

You won't often see a big yellowmargin moray out cruising solo, but she's a beauty to behold when you do.

But the reef is not all eat or be eaten. In any society, a few hearty individuals underscore the *joie de vivre* awaiting discovery on a reef not ravaged by the aquarium trade.

Yellowtail coris seeks friendship & possible LTR. And what do you know, a response:

I like to swim in the moonlight and quiet evenings on the reef, so... Yes! I say yes. Again yes.

And in the luxury of nature untrammeled, baby makes three:

Or four:

Yellowtail coris juvies
What's the difference between "sustainability" and fish in abundance? Many fish.

Among the coris clan is the elegant coris, common to protected reefs.

It's easy to hang in a neighborhood like the north edge of the sand channel at Molokini Crater, with its abundance of fish engaged in fluid society with no fear of unnatural death and merchants of "sustainability."

Heading down the channel a step and a stumble, such as it is, we come to another ethnic section, home of a territorial, feisty clan, the Hawaiian dascyllus—damsel fish who draw into reef cover on the approach of any big swimmer—then ease back out if the swimmer bears no bad intention.

How do I, Snorkel Bob, mystically capture the submarine porthole effect? EZ! You just keep your telephoto lens where you can select it by mistake…*et voila!*

Up close, the dascyllus will often pass in curiosity…

…or step up in defense of a home reef, making him one of us.

Like all reef fish, dascyllus love a cleaning—IF the Hawaiian cleaner wrasses have not been pillaged for the aquarium trade, which sorely compromises the reef and dooms the fish.

Once Cleanerboy goes to work, the fish line up for parasite removal. The goldring surgeon is an herbivore, vital to reef health, who in turn will clean a turtle or two on his daily rounds.

Goldring surgeon, or Kole, all buffed out.

A cruise across the sand channel to the south edge takes in the home turf of a few triggerfish, like these pinktail durgons…

...or this ornate triggerfish.

photo by Anita

Triton triggerfish

Just for fun, here are some triggerfish cozzins, 1st a triton trigger on a Great Reef bommie. 2nd, a yellowmargin trigger at the Sofitel sand flats on Moorea, in French Polynesia. Differing complexion, gill plates, and certain other *je ne sais quoi* cannot hide the family resemblance:

Yellowmargin triggerfish

photo by Anita

The bluegill or gilded triggerfish buck is flamboyant and yellow-tailed, while the doe is coy and flirtatious (top left).

At the south edge and on across Middle Reef are resident Potter's angelfish—protected and willing to pose every now and then.

Fish depend on coral for sustenance and shelter. While the angelfish might come up to look around, squirrelfish rarely swim squarely onto center stage. *Spotfin squirrelfish (above)*

Hawaiian squirrelfish

Bigscale soldierfish

Blushingly tentative, soldierfish linger under ledges or other cover. He's social by nature but prefers his own ilk.

Pearly soldierfish

Sometimes a soldierfish grants a mug shot, up close & personal, and it's a ringer for…

Uh…fool me once, and uh…shame on…uh…

Remember that story?

A cruising puffer most often turns away as if self-conscious, but sometimes he'll turn back for an eye-to-eye, to see who and what you be.

Sadly, this is the face of *fugu*, what the Japanese eat for the exhilaration of cheating death, though the puffer dies tortuously in the process while showing far more soul than the murmuring patrons in the background—that's on the YouTube horror show of *fugu* prep. I, Snorkel Bob, won't buy a Honda or a Toyota, but here I am shooting a camera made in Japan. What can we do? I don't know, but I suspect a correction will come. Meanwhile, anxiety abounds on the southern edge of the sand channel at Molokini Crate.

I'm not depressed, really, I just can't fathom how they can justify…

Just overhead, most often schooled up for a cruise but occasionally coming close in curiosity, a pennant butterflyfish scans the action.

Abundant at Molokini Crater in recent years is the blacklip butterfly, a smudged version of the milletseed till you get close, either among the coral…

…or in the water column, mixing freely with the free swimmers there, where his beauty is revealed in detail.

Remember my contention that familiarity engenders society? Molokini Crater supports the theory. But every waterlogged reef addict has to dry out sometime, and so we rise—wait a minute. A little coral tower just yonder has an unlikely orange speck in the center…

What ho! It's a flanneljammie blenny, a fish I'd call circumspect rather than shy. He won't come forth to check you out like some—in fact he'll skedaddle in a blink, unless he has eggs to tend. Then it's him or the Ms. front and center, standing guard.

Flanneljammie blenny can cast a wary eye…
or relax in full repose.
Ms. Flanneljammies is less strident in posture and coloration and may prefer no human contact. Here she is peeking out of her personal chamber, apparently interrupted while in dispose.

And on this note we hele on to another note of slightly higher (shrill again) timbre, 100 miles south to the North Kona Coast of the Big Island of Hawaii, to cruise shallow at Puako.

Puako is not protected. The flanneljammies femme here came out of nowhere to perch on some finger coral, her full flair pose apparently triggered by sudden awareness of my presence. She froze—click—and then vanished.

Do I, Snorkel Bob, blame her paranoia on the aquarium trade? The aquarium hunters would be disappointed if I didn't. And I do. Puako in its splendor is a reef intensely oppressed by the aquarium trade. The Hawaii State Division of Aquatic Resources Big Island Head Hotdog is a most ardent defender of the aquarium trade. Did I mention no limits on the catch or the number of catchers and no constraint on endemic or rare species?

The Kona DAR Head Hotdog defers to data, often swelling and, if the heat is up, sweating, as he transits freely from "science" to "economics." Empirically clear to him is that "the aquarium trade is our most lucrative near shore fishery."

Otherwise clear is more swelling, transforming the Head Hotdog to…the Big Bratwurst. This pattern of legacy-building within the scientific/bureaucratic matrix may be recurring. But he and I digress.

Reported aquarium catch revenue of $2 million annual wouldn't make the down stroke on a modest beach bungalow here in Paradise. Then again, the aquarium "fishery" does not report what it poaches. If you come to Hawaii, ask around; aquarium collecting is killing reefs and tourism, which, oh, by the way, is worth $800 million annual.

Kona DAR reported recently on aquarium hunting on two reefs—Honaunau, also known as the City of

Refuge, and Puako. The blueline butterflyfish has vanished from both reefs. In plain English with no smoke up the sphincteroo, the blueline butterfly is gone. Yet the DAR study—one of numerous studies now collecting dust—says the blueline butterfly is "experiencing a 100% decline" at Honaunau and "98% decline at Puako".

We interrupt this program to say

BAH!

Back at Puako, we'll take a minute in memoriam:

Where are these bluelines? They're not on the Kona Coast, where the State "manages" a free-for-all gold rush, and the gold nuggets are reef fish. The situation is grim.

Kona DAR was caught in the middle of a range war—commonly called the Aquarium War—complete with live ammo fired from the shoreline on the reef pirates, who fired back.

Like Alec Guinness before him, the Kona DAR Head Hotdog gained a ceasefire by compromising with the enemy to build the bridge on the River Kwai—an alliance that cannot last. A classic example of the moral vacuum in Kona is the fish kill at Honokohau boat harbor recently. A guy from Montana got wind of the easy money and no limits on aquarium collecting in Hawaii, so he came on over and dove right in. The hard reality is that holding tanks have no margin for error on chemistry—even if done right, the tanks must "run in" to get rid of toxins. So the guy collected 675 fish, mostly yellow tangs, and his tank failed.

They died.

Aw, shucks. So he bagged and froze them to avoid the stink, and a few months later dropped them in a dumpster at the harbor while running errands. The bag stunk, and by the Grace of Neptune, the bag was clear, not green. Somebody pulled it out and spread the fish out. Then came a hue and cry from the vast majority of human residents who hate the aquarium trade.

I mentioned earlier that this event was not isolated, even though the Kona DAR talking-head departed from reality-based data to announce: "This doesn't happen." The legal crux here is that the State called for an investigation by DAR, because DAR manages the aquarium trade as a "fishery." That management is in violation of State and Federal laws on animal abuse to protect animals raised or caught for the pet trade, which are far different than laws for food animals.

The other crux, larger than legality in its arrogant disregard of public sentiment, was the statement from Kona DAR, that yes, we have some moral questions here, but we're not here to address those—and this is our most lucrative near shore fishery, yadda, yadda, yadda.

Oh, a change is gonna come.

Meanwhile, back at Puako, we have a most expansive reef that is home to many fish as long as they can recognize a lowbrow with a net out to catch a heritage and sell it.

Among the survivors to date, a teardrop butterfly—**down 90% at Puako and Hononaunau**.

While all butterflies are beautiful, the forceps has notable flourish:

Forceps butterflyfish

Puako reef also provides an excellent opportunity for those of you with degrees to use them. Can you tell what makes a longnose butterfly below different from the forceps butterfly on the facing page?

Longnose butterflyfish

Yes! A+

The longnose butterfly has a longer nose. Okay, follow up: can you tell the distinguishing characteristic of a zebra moray?

Zebra moray

If you said stripes, you may be ready for grad school. Which just about wraps it up for the Kona Coast, a most incredible reef and scene of a State-sanctioned degradation of a fragile natural system.

Which is hardly a good reason to visit, but in case you choose to come anyway, you may want to take a good look at this chevron tang, which is actually the juvenile phase of the black surgeon. You won't likely see this fish in Kona, but you can buy him online for $150.

Okay—here's a bright spot, maybe…

It's a featherduster worm who bores into a coral head and then sticks his duster out to filter-feed. Featherdusters got wiped out on Oahu, mostly in Kane'ohe Bay. How do you "collect" a featherduster? You smash the holy living snot outa the coral head to break the worm free. The aquarium scourge response: "But we don't take feather dusters anymore!" Because they're gone.

Wai wai wai wai wait, Snorkel Bob! You said this would be a brighter spot!

I did? Oh! Yeah. Well, you may still see a featherduster at Puako.

Which makes me, Snorkel Bob, feel good, as I do every day spent on a reef trying to recover from the ravages of "commerce." Like the reef at…

UKUMEHAME

between Olowalu and the pali on Maui, where many acres of finger coral should be prime nursery for yellow tangs and other small fry…

Go snake-eyed to check out the tiny goby on the finger coral stalk. This is interspecies communion and life. Nothing needs to be taken here for commerce to thrive.

You may see recovery as well at…

Kahili Reef

on Maui is a favorite habitat for me, Snorkel Bob, and many lei triggerfish, or humuhumulei.

Mmm...

Oh…

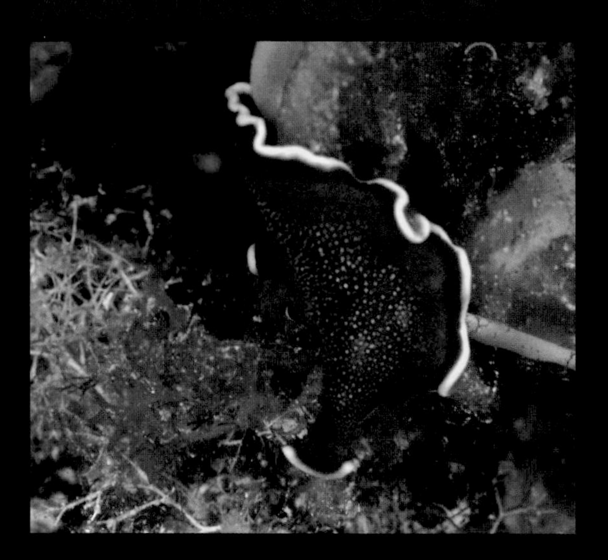

on Maui's east shore is a tsunami zone with a low lie that makes for pounding surf most of the year and keeps waterdogs out. Summer calm allows access to some of Maui's most undisturbed fishes.

Flatworms like to hang at Maliko, like this dime-size dazzler…

Maliko could be a fang blenny mating area. These little rascals are cute as buttons, till they take a tiny chunk from your hide; then you want to slap their adorable peduncles. The fang tango, however, is a rare dance to brighten a snorkel curmudgeon's day. Will fang blenny romance yield more tiny fang babies?

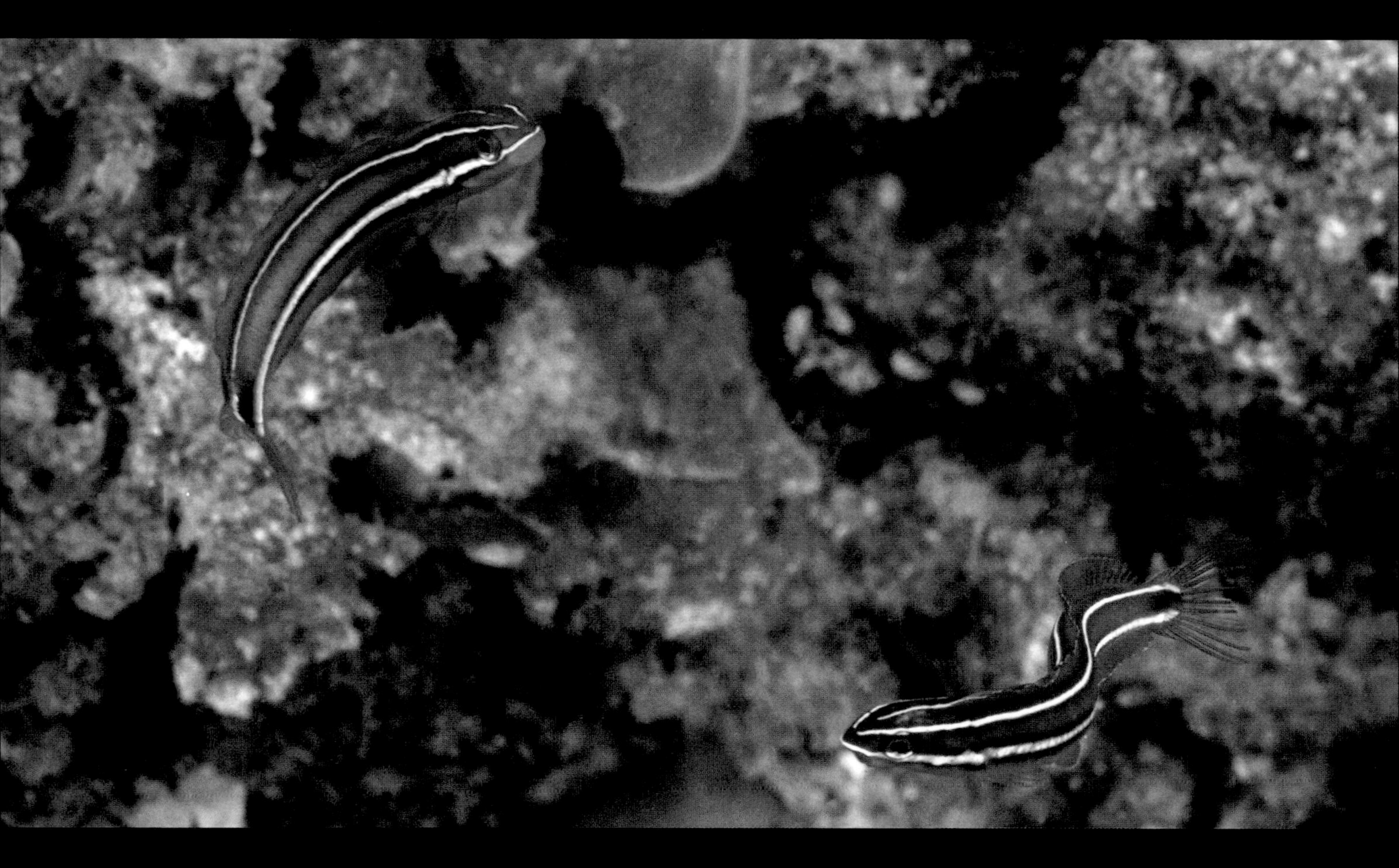

Fang blenny went a courtin' & he did ride uh huh. Uh huh…

Wait a minute: do fang blennies fang each other in a horrific rendition of blenny S & M? Nah! Couldn't be!

This is not a lost garter but an egg case, likely deposited by a flatworm or Spanish dancer.

Privacy seems to spawn romance at Maliko Gulch, or maybe in springtime a young multiband butterfly's heart naturally turns to love.

LANAI

is where fewer people mean less effluent and more fish.

It's 45 minutes on the water taxi out of Lahaina Harbor, and the fishy character of Lanai is mostly because of fewer fishermen—and fewer aquarium hunters. The good news is that Lanai is part of Maui County, the most progressive Hawaiian Island in seeing a severe threat to the public welfare for what it is and moving to correct the problem. What's the difference to date between Maui County and the State of Hawaii? What's the difference between a vision and standard operation? It's politics, which comes down to personnel.

These two frogfish are the same species and share skills in color changes. Sitting mere paces apart, they may be on the verge of romance too, unless they're immersed in oblivion. We can only speculate, though a most notable behavior of Commerson's frogfish is attempting to eat prey bigger than itself, so we generally observe them from beyond gulping distance. How embarrassing, returning with a frogfish enveloping yo head (!).

Lanai is known for its caverns, the roofs of which may host swimp, I mean shrimp, like these fountain shrimp, so called because of their antennae arcing like fountains.

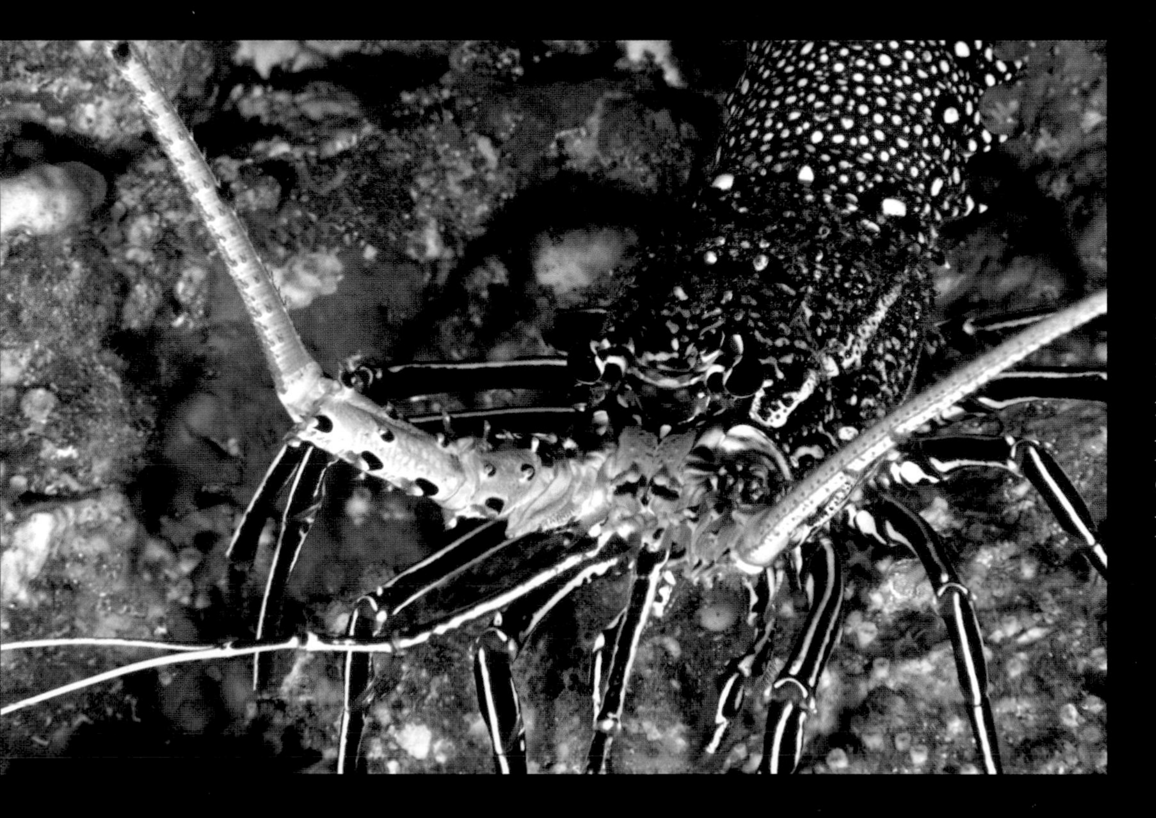

This roof dwelling lobster was not seen in Hawaii till 2009—note the unique pink piping. The underside of the tail is matching pink!

The raccoon butterflies are still prolific on Lanai, and they're one of the butterfly species fond of schooling.

A smaller cavern yielded a blushing trumpet. Trumpetfish change color to blend with other species, so they can hunt undetected. This particular trumpet on Lanai, however… …turns red only at Chanuka, so he can guide Neptune's sea sleigh as it rounds the reefs of the globe on cloudy nights, delivering toys and gelt to young reef fish who must turn in early or he won't come. It's true. I, Snorkel Bob, did not make that up.

These young puffers today; what's a snorkel executive to do? Okay, here's our last story for a while, and it's shaping up for a happy ending. Who knows what a few key humans might do, but we can hope for the best. You'll have to cut me some slack here, boys and girls, because this is a reptile story, not a fish story. But these sea beauties can hardly be counted out of what every reef addict loves. Are you getting my drift? It's…

The Turtle Formerly Known as Prince is a critically endangered hawksbill on the reef fronting the hotel formerly known as the Prince, South Maui

TURTLE TIME

in which I, Snorkel Bob'm pleased to announce that

A NON-PROFIT ORGANIZATION DEDICATED TO REEF DEFENSE

is underwriting litigation by Earthjustice, the nonprofit, public-interest law firm, in Federal District Court against the National Marine Fisheries Service for allowing expansion of the Hawaii longline swordfish fishery and a dramatic increase in allowable take of threatened and endangered sea turtles.

Hawksbills were hunted to near extinction for their red shells, for mirror handles, eyeglass frames, and other chachkas. Hawksbill turtles feed on small sponges and would like to make a comeback on their wilderness nesting beaches but won't if they continue to die on longlines.

This case highlights outdated ocean management policy that must change to allow for ocean recovery. The oceans can no longer provide limitless protein for growing human populations. The National Marine Fisheries Service (NMFS) is now part of the National Oceanic & Atmospheric Association (NOAA), a bureau in the Department of Commerce. That means ocean management policy is based on revenue, and not on recovery. These agencies should be in the Department of the Interior, where conservation and recovery are management priorities.

The Western Pacific Regional Fishery Management Council (WESPAC) recommended the swordfish fishery expansion. WESPAC uses the word "conservation" more than most agencies, but its members and officers are vested in commercial extraction. Efforts to override the evidence and global consensus are not new, and the bureaucracy grinds slowly. Like a giant ocean liner, the Commerce Department changes course slowly. With an administration agenda reflecting long-term ocean recovery, we go now to the judicial arena, with Earthjustice seeking exactly what it's named for. I, Snorkel Bob, call this money well spent.

Green sea turtles (*honu*) and hawksbill turtles (*honu 'ea*) are often seen today. Hawaiian waters extend for hundreds of miles and include migratory routes of loggerhead and giant Pacific leatherback turtles. Leatherbacks grow to 8' in shell length and cruise with an entourage of small sharks, remora and others in a procession to make Neptune proud.

Loggerheads in Hawaiian waters all too often snag a meal with a longline hook inside, so the turtles drown. Loggerheads were listed as endangered in 1970 and have lost significant nesting area since then. Primary migrations are east/west both above and below the equator. The ancient mariner on the next page is settled for now into community life on the Great Reef.

Casual as any being unburdened by fear, this huge loggerhead cruised and nibbled, another face in the turtle crowd, till he drifted curiously close to a mermaid with a camera for the personal portrait.

Do you think this turtle has logged many sea miles? These turtles too often fall victim to longlines and die for the swordfish special. Besides speaking up on aquaria, it's time to stop eating swordfish. The mercury content will kill you for starters, leaving the world without three personalities worth saving—the fish, the turtle and thee.

But enough of the dark cloud. Let us remember to defend what we love and have a good time too. Because fade as it may, nature should not fade away as we sit mum. On that note, I give you a reef stalwart who will stand-up, step up and speak up for all things turtlistic…

…Honuanita, main mermaid to me, Snorkel Bob. All these turtle portraits in her family album are snorkel shots, free dive encounters with turtles who trust humans.

Honu is Hawaiian for green sea turtles, so called because of their vegetarian diet that renders their "meat" green. As recently as 1969, turtles alive and dead were tossed into waiting pickup trucks at the fuel dock at Lahaina Harbor. *Honu* were eaten to near extinction. Now common in near shore waters, they often engage snorkelers and divers, sometimes with affection.

Honu 'ea (ay-uh) is Hawaiian for red turtle, or hawksbill. The hawksbill diet of sponges made them carriers of ciguatera, a toxin accumulating up the reef food chain to the apex predators. Ciguatera is harmless to reef animals, though it causes humans who eat those animals to itch for two years and then die, a tough break indeed.

A few familiar *honu* are…

Grumpy lives at Oneuli (black sand) reef at the base of Pu'u Olai (Red Hill). He tips in around 300 lbs.

Son of Grumpy lives on the reef next door, at the hotel formerly known as Prince.

Honu, or Hawaiian green sea turtles, relaxing at the resort…

Hawksbills are easy to distinguish by the hawk-like schnoz and a more dramatically serrated back edge of the shell. The flippers also have a more pronounced skin pattern. While *honu* can grow to 450 lbs, *honu ʻea* are smaller, rarely topping 300—though rareness in this application may be redundant.

Here again is The Turtle Formerly Known as Prince, who tipped in at 142 lbs and growing.

Hope is a good bit larger, tipping in at 165 lbs. Though solitary creatures, Hope occasionally cruises and feeds with The Turtle Formerly Known as Prince.

All turtles work in a surge sooner or later, this way and that while grazing or turning rubble in search of sponges. This young hawksbill is 'Oli, curious as any young turtle would be. Checking out this action requires fending off the coral with a front flipper.

I, Snorkel Bob, urge you to visit your nearest reef and show your love. I started humbly with a point 'n shoot. Years and thousand$ later—the boat, the camera, housing, strobes, arms, lenses, ports and stuff (the frikkin synch chord was $250!) I see the artistry available with mask, fins, snorkel, and another point 'n shoot, and I am further humbled:

(NOTE: *Hawksbill Babies at Oneloa*, the movie, is still free to educators at snorkelbob.com.)

It's 'Oli again, surfacing for a breath and to better show the algae growth on the shell. 'Oli is estimated at 6 years old here—and in this light shows another fundamental difference between honu and hone 'ea. It's in the scutes, or sections of shell. Honu scutes are mosaic, like tiles on an even plane with little grooves separating. Honu 'ea scutes overlap like roof shingles, which may have caused the desecration to hawksbills, by humans destroying them to make novelty items from their shells. INSET: Hawksbill babies at Oneloa.

How are the weights of these turtles known? Because a fellow from the National Marine Fisheries Service had not seen a hawksbill live on a reef in decades, so he called Honuanita for guidance. She showed the way to favorite watering holes of three friends on two reefs in short order—so the NMFS fellow could catch and weigh them.

Ah, for science.

The Maui Snooze made vague reference to a snorkeler who went along to "assist." This is the same journal profiled in *We'll Always Have Chicago*, a story of turtle love from the collection *Wintner's Reserve*, available at amazon.com or snorkelbob.com.

Also notable is that this NMFS scientist was instrumental in listing honu as endangered many years ago but now supports the WESPAC move to de-list honu. He denies speculation that he wants to secure his "legacy" by completing the cycle. But the Hawaiian word for this move is still humbug. Please, don't get me, Snorkel Bob, started on greed and/or self-aggrandizement. The Hawaiian Islands now host a new social order that some people of the human and turtle varieties hold dear. Don't worry, old Data Dog; we won't forget you or your good deeds, like when 'Oli the young hawksbill developed a lump on her neck.

Honuanita observed its growth on a few visits and called the urban core. The science crew returned to capture 'Oli for diagnosis—not a tumor but an abscess—and veterinary treatment by lancing and antibiotic flush for a few days.

Natchurlly, a turtle lass can be spooked by abduction and intrusion, but in a short while she eased back to acceptance that some humans are driven by love, only love, in a place of unfathomable beauty, of social order and friendship.

Which isn't to suggest that friendship or society is always easy. Like just this morning when, to Surlygirl's consternation, Peewee arrived.

Uh…I was thinking we could…uh…

Beat it, ya mutt!

A reef balances habitat among inhabitants the same way some humans do, through rational discourse and a settling process. Do we have an understanding?

On the first pages of *Some Fishes I Have Known*, some inveterate reef dogs searched for the Holy Grail—the Hawaiian seahorse seldom referenced and yet to be seen. That pursuit proved yet again that you can't always get what you want, but if you try sometime, you just might find… Revealed when least expected:

The Next Dalai Lama

Many moons later, again in the shallows, another Stones ballad bubbled up as if at random.

So take me to the airport.
And put me on a plane.
I've got no expectations, to pass
through here
again…

When there before us, her tail wrapping the base of a halameda stalk: the Golden Girl...

Hawaiian seahorse males are deep green, the femmes golden. Rolling her eye with questions and answers she waited in ultimate grace.

Oh, seine trawlers drag seahorses from the bottom every day; dried seahorses sit in souvenir shops around the world. But that's easy. That's humanity killing nature, as it will likely do till it kills itself. In the meantime, those needing hope and a reminder of what endures may find solace in this everlasting light.

That's all for now, boys and girls. Stay tuned to find out how you can jump in to help keep reef fish at home where they belong—on the reef!

Remember: Aloha is more than a greeting or a farewell. It's a way of life. Aloha has many uses and interpretations. Among them is an original version of the golden rule, directing us to *take care of each other*. This directive is not limited to human people but to people of all species in the land around us.

A Hui Hou

Reader's Service Page

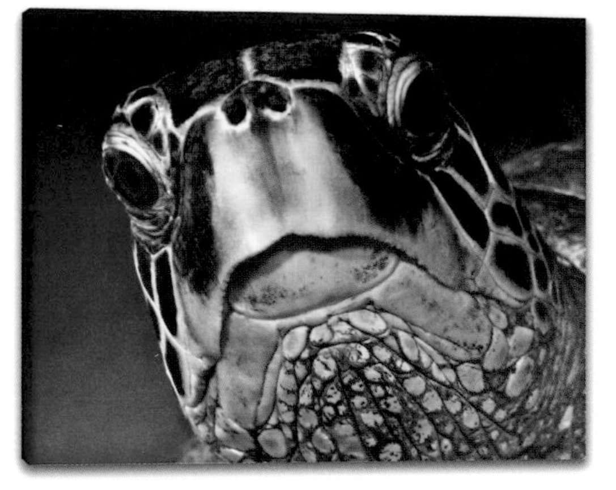

Reef Art?

You can order a fine art giclée on canvas of any photo(s) shown here at (800) 262-7721 or snorkelbob.com. Specify 17", 24", 36", or 44" width. Your gallery-wrap canvas will arrive prepped for stretching.

Reef Time Hawaii is 35-minutes of finned frolic filmed on location, with Snorkel Bob and longtime snorkel buddy Matt Roving. The Reef Time DVD includes the 22-minute epic, *Hawksbill Babies at Oneloa*, as a bonus track. You. Will. Love this rare footage of hatchling turtles slogging to the sea at Makena Beach, Maui. With music by Keali'i Reichel, the Ka'au Crater Boys & Wayne Perry on ukulele you'll drift again to the balmy shores.

About the Author

Robert Wintner is the Executive Director of the Snorkel Bob Foundation and works diligently to protect Hawaii reefs. He led legislative campaigns at the Hawaii State Capitol and in Maui County from 2007 through 2011. First passing the State Senate with unanimous consent, the aquarium campaign got derailed in the House by vested interests in leadership and the Executive Branch. These efforts brought aquarium plunder into the light of day, however, with a grim reality and empty reefs facing the people of Hawaii. Legislative efforts will continue.

Wintner's short fiction has appeared in Hawaii Review, (U. of Hawaii) and Sports Illustrated. His 2007 novel, *In a Sweet Magnolia Time* was nominated for a Pulitzer Prize. Wintner's adventure novel of the sailing charter trade *Whirlaway* and *Modern Outlaws* were both optioned for film rights by Hollywood production companies. Other novels and accolades over the years led to *Some Fishes I Have Known*, a first glimpse at the social side of the reef community.

from left: Lulu and Robert Wintner

Every Fish Tells a Story brings into focus some key personalities from reefs across the western world, along with insight on the devastating aquarium trade. Asked in a radio interview if a fish can have a soul, Wintner said, "I wonder if some people have souls. The question of souls and who might have one is strictly human. It's a luxury, a mind game, and it's demeaning in the sense that fundamentalists can deny souls to some species. Humans tend to define intelligence in other species as a measure of the ability to learn certain behaviors. But many species decline to learn those behaviors and still have lessons to teach—to those of us willing to learn. So let me rephrase the question: 'Can a fish be a friend of mine?' The answer is yes, obviously. Look at these pages."

Wintner is still a voice for marine mammals, turtles, the reefs and its citizens, and vows to continue. "It's a long march. We have no choice but to take the next step."

Index

U

V

W

Y

Z